W. H Spackman

Trout in New Zealand

Where to go and how to catch them

W. H Spackman

Trout in New Zealand
Where to go and how to catch them

ISBN/EAN: 9783744793995

Printed in Europe, USA, Canada, Australia, Japan

Cover: Foto ©berggeist007 / pixelio.de

More available books at **www.hansebooks.com**

TROUT IN NEW ZEALAND:

WHERE TO GO AND HOW TO CATCH THEM.

BY

W. H. SPACKMAN, B.A.,
BARRISTER AT LAW,

President of the Canterbury Anglers' Society, and Counsel to the Canterbury Acclimatisation Society.

WELLINGTON.
BY AUTHORITY: GEORGE DIDSBURY, GOVERNMENT PRINTER.
—
1892.

PREFACE.

THIS work has been written for the purpose of supplying a want long felt both by anglers in the colony and visitors, who are at a loss to know where to go and what to use in order to obtain the sport which is to be found with trout in New Zealand. Most anglers, even in New Zealand, are unacquainted with any district but their own, and strangers to the colony have had no means up till now of obtaining information except from personal inquiry from the friends they may chance to meet in their travels. It is hoped that this little book will give them most of the information they may require.

With regard to Canterbury I have had no difficulty, being personally acquainted with the whole of it. Want of leisure-time, however, prevented me from obtaining that local knowledge myself of the other districts, without which this work would be of little use to the angler and tourist. Had it not been for the assistance given me by the acclimatisation societies, and by friends living in the various parts of the colony, I could not have written it. To them I desire to return my sincere thanks. A few, however, have given me so much substantial assistance that I must make special mention of them. The chief of these are my friends Mr. A. J. Rutherford, of Wellington; Mr. Kingdon, of Nelson; Mr. A. M. Clarke, of Temuka; Mr. Knubley, of Timaru; Mr. D. Russell, Mr. Begg, and Mr. W. S. Pillans, of Otago; and Mr. Edward Tanner, Mr.

Russell, Mr. Feldwick, Mr. Campbell, and Mr. Handyside, of Southland. Last, but not least, I desire to acknowledge my indebtedness to the valuable papers, in the "Transactions of the New Zealand Institute," by the late Mr. Arthur, one of the fathers of acclimatisation in New Zealand, a man of great ability and original research, and whose death some years ago is still widely lamented and deplored.

W. H. SPACKMAN.

Christchurch, September, 1892.

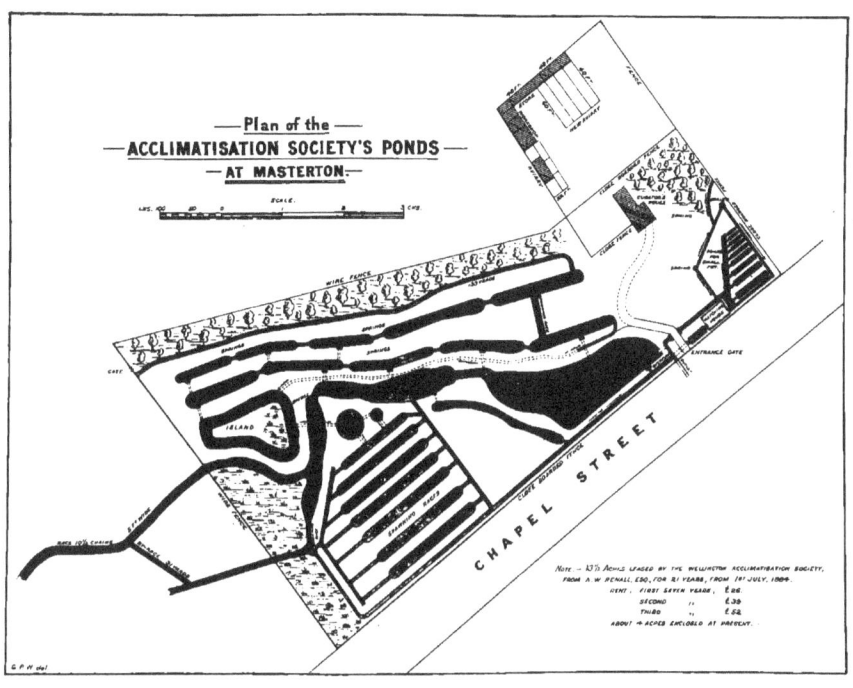

TROUT-FISHING IN NEW ZEALAND.

CHAPTER I.

Acclimatisation in New Zealand—First Beginnings in Otago and Canterbury—The Hatcheries at Clinton and Wellington—Different Times required for hatching out Ova—Salmon Experiments—Examination of Causes of its Success and Failure.

SEVEN great rivers—the Clarence, the Waiau, the Waimakariri, the Rakaia, the Rangitata, the Waitaki, and the Clutha—find their exit into the sea on the east coast of the Southern Island of New Zealand. They are cradled in the great southern range, the Southern Alps, which form the backbone of the South Island. They are all known as snow-rivers—that is, rivers fed by the melting of the snows in the great range. They are at their lowest in the winter, at their highest in the spring. The beds of some of them—the Waimakariri, the Rakaia, and Waitaki—are in many places more than two miles across; not one of them, except the Clutha partly, is navigable. When the snows melt, in the spring and early summer months, they become huge resistless torrents, their beds filled with muddy or milky-coloured water, carrying from the mountains thousands of tons of shingle to the sea.

There are a few smaller ones known also as snow-rivers: such are the Hurunui and Ashburton.

But there is another class of rivers in the South Island, known as rain-rivers—that is, rivers draining small portions of the country, and whose sources, not extending far enough back into the ranges, are unaffected by the melting of the snows. These are very numerous, and in anglers' parlance they are known as rain-rivers.

Twenty-five years ago not one of these rivers had the least interest for the angler; they flowed through some of the most beautiful gorges in the world; they travelled their courses over the great plains unheeded and unloved. None but the dweller on their banks cared the least in what condition were their waters. The bullhead (*Eleotris gobioides*) lived beneath their stones in thousands; the young of the sea-mullet and the inanga ascended them annually in millions; naught but enormous eels and useless shags preyed upon these millions of fry. The settler cursed them for encroaching on his land; but the rod of the fisherman never cast a shadow on their waters; every one of these mighty rivers, every one of the thousand creeks and streams that flow into them, or that water the islands of New Zealand, were tenantless and profitless to the sportsman.

What a change there is to-day! From its head-waters, far beyond Lake Tekapo, Lake Ohau, and Lake Pukaki, to the sea, the Waitaki, next to the Clutha, the greatest of these great snow-rivers, is the home of the English trout—trout of such a size, too, that accounts sent to England of their size and of the sport obtained with them were long looked upon as fabulous, and received with derision and incredulity. From the Waitaki, from the Rakaia, from the Waimakariri, from the Rangitata, has the angler taken more than 100lb. weight of trout in a single day with the Devon or the phantom. Often has the fly-fisherman captured more than 30lb. weight of trout with the Governor or the Moorfowl, the Soldier-palmer, or the March brown. A 10lb. trout from any of the snow-rivers is a thing of every-day occurrence, and they have been taken up to 25lb. With the fly trout have been caught exceeding 8lb.

The history of the acclimatisation of trout in New Zealand is the history of an astounding success. It began in Otago in 1867, and the first successful hatching of trout there was in October, 1868. This was achieved by Mr. Clifford, the curator to the acclimatisation society, who went to Tasmania and got from the spawning-beds at the breeding-ponds of the Plenty 800 ova: 720 of these were hatched out. Part of this lot was sent to Lake Wakatipu, but all the young fish died on the road. The remainder were turned out at Palmerston, in the mill-race on the property of a Mr. Young. A year afterwards one of these fish was caught and found to be 7in. long. In October, 1869, the second shipment of 1,000 trout-ova was brought from Tasmania by Mr. Clifford. The fish from these two lots of trout-ova form the original stock which were liberated in Otago streams in November, 1869, and from these and their descendants the ova for stocking the rivers in Otago have been obtained. Between 1869 and 1891 nearly two millions of brown trout alone have been hatched out by the Otago society, and distributed in upwards of three hundred streams.

The hatchery at Clinton, in Otago, is now the largest in the colony, and was first established in 1885. It is supplied from a source called Marshall's Creek, a mountain-stream and tributary of the Kuriwao, the average temperature of the water during the hatching period being 41° Fahr. The two hatching-houses have room for hatching out nearly 700,000 ova, and the number annually distributed exceeds 250,000. About 50 acres have been enclosed by a wire and standard fence. The period for hatching-out is found to be, at this temperature, for brown trout seventy-five days, for American brook char (Fontinalis) and for Scotch Burn trout seventy-four days, and for Loch Leven trout seventy-one. Besides a number of ponds and races for the small fry, there are three large ones. No brown trout are now kept in stock, as large quantities of ova of brown trout can be easily obtained in the months of June and July from the Waiwera, Wairuna, and Kuriwao, but about fifty American brook char, 500 Loch Levens, and 220 Scotch Burn trout are kept in the ponds for the purpose of obtaining the yearly supply of eggs.

Beyond the fact that our trout are believed to be from a Thames tributary, little is known about them. As mentioned above, the original stock in Otago was obtained from Tasmania, and it appears that three lots were sent to Tasmania which turned out more or less successful. Of these, Mr. Francis Francis sent one from the Weycombe, Bucks, and another from the Wey at Alston, Hants; and Mr. Buckland sent one lot from Alresford, on the Itchen, Hants. From one or other of these sources most of the streams in Otago were stocked.

In Canterbury, in 1866, ponds were formed in a piece of ground forming part of the Public Domain at Christchurch. These ponds were at first supplied with water from a stream running through the south of Hagley Park which empties itself near the Hospital. This not being found very suitable, an artesian well was put down, which yielded, and has continued to yield ever since, an abundant supply of pure water, free from sediment, at a low and even temperature—namely, 53° Fahr. Other wells were afterwards sunk, and a copious supply obtained from them. Though deficient in food, this splendid supply of pure water, always at the same temperature, has been one of the reasons of the great success attained in the breeding of fish in Canterbury.

In June, 1867, the curator proceeded to Hobart and obtained 800 ova, but, owing to a long passage, most of them were destroyed, and only three hatched out. These, unfortunately, escaped. In 1868, however, a substantial success was obtained: a further supply arrived from Tasmania through the Otago Provincial Government, a large proportion hatched out, and 433 young trout turned out in various places, besides 112 being retained in the society's ponds as a reserve for breeding. In the year 1869 another lot was obtained from Tasmania *viâ* Dunedin: from these, 450 were obtained, but most of them escaped—probably into the River Avon. In 1870 ova was first taken from the trout in the Gardens, and 292 were sold at £2 per dozen and liberated in various streams in the province. Since then, every year up to the present time, the number of trout hatched out has been increased, and very many thousands distributed, being either purchased from the society by private individuals or liberated gratis, so that now almost every little rivulet has received some fish. Yet much remains to be done in the back country both in Otago and Canterbury. Such a success, however, as this has been could not have come about without great enthusiasm. It can be imagined that men who would give £2 a dozen for little fry, only about an inch long, would undergo much rather than lose them in transit. One enthusiast, who obtained fifty in 1870, carried them in a bait-can in front of him on his horse for sixty miles, swimming a river or two *en route*.

In the Wellington Province the first attempt to introduce trout seems to have been made in the Wairarapa district by private individuals. In 1874, Messrs. Beetham and Rutherford purchased fifty young trout from the Canterbury Society, and liberated them in the Waipoua River, but nothing apparently ever came of them. In 1876, Messrs. Beetham Brothers obtained about 2,000 trout, which were hatched out at Brancepeth and turned out in the Wainuioru and Kaiwhata Streams. From time to time an occasional trout was seen; but little result had been obtained, the streams not being very suitable. Several tried after this to introduce trout, but they never seemed to increase. It was evident that if the thing was to be a success it must be undertaken systematically. In 1882, an acclimatisation society was organized for Wairarapa, which in 1884 amalgamated with the Wellington society. A hatchery was established at Masterton, and 10,000 ova of the brown trout, obtained from Canterbury, successfully hatched out and liberated. In 1883, 15,000 were obtained through the courtesy of Mr. Arthur, the Chief Surveyor of Otago; but, after about 3,700 fish had been liberated, the rest were stolen from the boxes at night by some miscreant, and one season was thus almost lost. It was, however, only a temporary check, as from that time onward very large numbers of various kinds of trout have been turned out every year.

The hatchery establishment at Masterton is one of the most complete and valuable in the colony, and from it trout have been sent to almost every part of the North Island and northern portions of the South Island. Its natural advantages are unsurpassed, and the arrangements are exceedingly good. The plan shows the general arrangement, its ponds being supplied with spring water, carried in races to every part of the grounds. About 4 acres have been enclosed, partly by a close-boarded fence and partly by wire. The hatchery is situated as near as possible to the curator's house, where it is easy of access, and can be visited in any weather and at any hour. It is most essential for the curator to be as near as possible to the hatchery. No one who has not had experience of fish-hatching can imagine the constant supervision necessary to prevent a loss of perhaps the whole of one season's supply. All intakes and outlets must be kept in the most perfect working-order, and all diseased or dead ova duly removed. It is almost necessary to have one person to look after all the water arrangements and another to look after that person. Many things were learnt in the early days of fish-hatching—how a whole season could be lost only through the want of a little extra care ; but the lesson has been valuable, and the experience gained has enabled what was but an experiment at first to be now a complete success.

From the ponds at Masterton in 1891 the society distributed nearly 350,000 fish, consisting chiefly of brown trout (*Salmo fario*), the American brook trout or char (*Salvelinus fontinalis*), and the Loch Leven trout. Many thousands were sent to a distance in large cans, containing from 500 to 1,000 each. This is the ordinary mode of transit. Last year some very interesting experiments were made in the transit of trout-ova and fry between this colony and New South Wales. Five thousand brown trout were sent to Sydney, where they arrived in perfect condition, and 1,000 Fontinalis fry, in the alevin stage, were safely carried in the ordinary cans without a change of water. A small number of feeding fry, hatched from Wellington trout-ova in Sydney, reached Wellington alive, a distance of 1,800 miles, having been sent from Sydney in hermetically-sealed jars, carried at a temperature of between 50° and 60° ; they lived for eight days from the time they were first imprisoned. The experiment was repeated from Wellington, and trout-fry, previously prepared by a day's starvation, were placed in hermetically-sealed glass jars, and reached Sydney alive, after being immured ten days three and a half hours of trying journey. partly in the steamer's ice-house and afterwards by rail, at a temperature varying from 53° to 63°.

In all, at the hatchery, besides the number of ponds for the small fry, there are upwards of twenty large ones for stock fish of different kinds, which are kept for breeding purposes. These are annually stripped and then returned to the water, none the worse for the operation. In 1886 the Government sent to California and procured ova of the rainbow trout (*Salmo iridcus*). This is an exceedingly valuable addition to the salmonoids, as its stands a much higher temperature than the ordinary brown trout, and is thus far more suitable for turning out in the more northern parts, especially round about Auckland. To show what loss a little carelessness will occasion:

during the season of 1891 about ninety rainbow trout, obtained from some ova received from the Auckland society, were put in a large rearing-box with wire-netting lids. These fish were growing splendidly, when some visitors to the ponds lifting one of the lids, and leaving it off, a large eel got into the box, and when discovered next morning its stomach was packed with nearly all these valuable fry, leaving only twelve alive in the box. The black Scottish Burn trout, perch (*Perca fluviatilis*), carpione trout, the English salmon (*Salmo salar*), and a few other representatives of the salmonoids can be seen in the ponds. Another is the native grayling, called by the Maoris upokororo (*Prototroctes oxyrhynchus*), one of the very few representatives of the salmonoids in the Southern Hemisphere. In 1888 a few of these graceful fish were captured, and placed in a pond, where they have become very tame. According to Dr. Günther, only two genera of these species are known—the haplochiton, abundant in lakes and streams emptying into the Straits of Magellan, also found in Chili and the Falkland Islands; and the New Zealand grayling. They are a mysterious fish, about whose habits little or nothing is known, found in one part of a stream to-day and disappearing no one knows where to-morrow. In the Wairarapa and some of the West Coast rivers—notably the Otaki, and also the Teremakau—they are found occasionally very numerous, and often run up to 2lb. in weight, giving capital sport and making an excellent table-fish; but the large trout seem too much for them in well-stocked streams, and they gradually disappear as the trout increase.

The period of hatching appears to vary very considerably according to the temperature of the water. In a report on the Crayfishery in Kent, England, the time is given at from seventy to eighty-four days. This is very much longer than that given by Yarrell on the result of an experiment in Germany, where it was found to be only thirty-five days. The Rev. W. Houghton, in "British Fresh-water Fishes," 1879, gives the average as sixty days, and the temperature 40° to 45°; and Mr. Francis Francis, in the "Practical Management of Fisheries," 1883, gives the mean time recorded as about sixty-three. The late Mr. Arthur, Chief Surveyor of Otago, gives the average, at the Opoho hatchery, of a series of observations extending over a number of years—1868, 1869, and 1878 to 1882—as seventy-eight days, the temperature varying from as low as 39° to as high as 57°. At Clinton, as mentioned before, the time for brown trout has been found to be, on an average, seventy-five days, with the temperature of the water 41°. In Canterbury the temperature of the artesian-water supply varies extremely little, being 53° Fahr., and the period of hatching of the brown trout is seldom more than from thirty to thirty-two days, an immense difference as compared with Otago. At the Wellington hatchery the temperature of the water varies from 50° to 56°, and the average time taken by the brown trout is thirty-five days.

As stated previously, the original ova from England were taken from three rivers—the Weycombe, Buckinghamshire, and the Wey, by Mr. Francis Francis; and the Itchen, by Mr. Frank Buckland. Taking Dr. Günther's arrangement of species, it determines the trout taken from any of the above-named rivers to be *Salmo fario ausonii*, or the southern form, probably the same as the Thames trout. Mr. Francis, in the work above referred to, tells us that Thames trout spawn early, or in November, and are correspondingly early in getting into condition in the following spring. This same trout in Otago spawns from the middle of June till the end of August, according to locality; and in Canterbury as early as the middle of May up till about the end of July, though occasional specimens are caught with fully-developed ova both considerably before and after these dates. Mr. Arthur concluded that the same species of trout in Otago is about two months later in spawning, and the ova ten days to a fortnight longer in hatching than in England. In Canterbury, however, the same species, appears to occupy only about half the English period of hatching, and to spawn at about the same time of the year, allowing for the change of seasons, as its English progenitor. Of the fish kept in confinement at the Masterton hatcheries, the Amerian brook char is always the first to spawn, commencing very regularly about the 12th May; next come the Loch Leven trout, and after them the common brown trout. The time for both the brook char and the Loch Leven trout, both at Masterton and in Canterbury, is about thirty or thirty-two days.

The acclimatisation of the English salmon (*Salmo salar*) has been long attempted, but as yet without any great measure of success. Many thousands, too, of the Californian salmon were imported, hatched out, and turned into the rivers, but no full-grown salmon have been caught. In 1885 Mr. Farr, for many years the indefatigable secretary of the Canterbury Acclimatisation Society, was sent by the Government to England to procure a shipment of English salmon-ova. He was successful in obtaining a large quantity, chiefly from the Tweed, and equally successful in conveying them safely to New Zealand. These were distributed to the various societies, and subsequently by them turned into numerous streams. In several cases the young fish were kept till they were 5in. or 6in. long, or until the appearance of the smolt-scales, before being liberated, but still no result was obtained so far as the actual catching of a veritable *Salmo salar*.

In the years 1886 and 1887 the Government imported a very large supply of ova. By the "Ionic," which left England in January, 1886, nine boxes, containing 200,000, were despatched. These were packed by Sir James Maitland himself, to whom and to Sir F. D. Bell the colony is much indebted for the trouble and the enthusiasm they showed in order to make the shipment a success. With the exception of one box of salmon and two boxes of trout, these were placed in an icehouse specially designed for the purpose, and a man put in charge for insuring a supply of ice to the trays. The single box of salmon- and the trout-ova were placed for experiment in the refrigerating-chamber.

On the arrival of the ship in Wellington the eight boxes of salmon-ova were found in very good condition, but those carried in the freezing-chamber were all dead. These eight boxes were divided amongst the various acclimatisation societies, and a large proportion successfully hatched.

In 1887 the experiment was continued on a still larger scale. By the "Kaikoura," "Doric," and "Tongariro," 590,000 English salmon, 30,000 *S. fontinalis*, 40,000 Loch Leven trout, 100,000 Rhine salmon, 25,000 Rhine brook trout, 25,000 Alpine char, and 25,000 Carpione trout were shipped. Of these the most successful were the English salmon, in some cases more than 70 per cent. being hatched out, which speaks volumes for the skill and care bestowed upon them by Sir James Maitland.

The experiment of placing a few thousands in each river having apparently failed, the fry were kept for twelve months in races, and then turned into the Aparima or Jacob's River, where there was nothing but a very few brown trout. This river flows into the sea through a wide estuary off the southeast coast of Southland, the temperature of the water, on account of a cold current, being probably lower there than at any other part of the coast. The advantage of this will be seen further on. Since then further shipments have arrived, and, after being hatched out and kept till the young fry were considered large enough to take care of themselves, they were turned into the same river. Some were kept in the hatchery ponds in Christchurch, and also at Opoho, and at Clinton, in Otago. In 1889 there were at the Opoho hatchery eight four-year-old salmon, and at Clinton 116 four-year-old. At the Canterbury hatchery more than 5,000 eggs were obtained from the salmon imported in 1885; and in 1891, from the stock in the ponds of the Otago society, 20,000 eggs were got and hatched out from salmon that had never left the gardens or been to sea.

These salmon—some of Mr. Farr's shipment—are now six years old, and, although they appear healthy, the experiment of propagating salmon from a stock kept in confinement has not been altogether successful. Although they have plenty of room and are well fed, and have produced ova for some years, they have not grown to their natural size, the largest being only about 2lb. The stripping of these fish is also a painful operation, the ova being equally as large as those of a large adult fish in its natural state; consequently, in the artificial operation of extracting the ova from the female, a large amount of pressure is necessary, which must injure the parent fish to a certain extent. An occasional fish has been found dead during the year, possibly the result of this treatment. The ova, however, hatch well, and the fry are always healthy.

In May, 1891, the estuary of the Aparima was netted and several fish taken. One of them was forwarded to Professor Hutton, of Christchurch, who has given much consideration to the subject, and who is recognised as one of the best authorities in the Southern Hemisphere. Professor Hutton was good enough, after he had examined it, to let me have it, and I obtained a trout from the acclimatisation gardens for the purpose of comparison. The conclusion we both came to was that the fish was a grilse. The reasons for the conclusions come to are too long to give here, but some of them will be found stated later on in the description of the Aparima River. This fish was taken in the estuary, and probably had never been further towards the sea.

A study of the ocean currents that sweep the coast of New Zealand will enable us to infer, from the experience gained in other parts of the world, the character of the marine life which stocks our seas. From observations, it appears that the coldest part of the sea round New Zealand is on the south and southeast coast of Otago, where the temperature of the surface-water ranges from 48° in winter to 57° in summer, the corresponding averages for the atmosphere being 43° to 58°. The cold current thus indicated, which probably exercises a good effect on the quality of the fish, besides limiting the range of a few species, appears to extend its influence up the east coast as far as Cook Strait, but on the west side of the Islands the average winter temperature of the sea is found to be much higher, and equal to that experienced 6° of latitude further to the north on the east coast. In the extreme south the summer temperature does not, however, rise to a corresponding extent; but, on the whole, there is evidence that the warm equatorial current which is known to skirt the east coast of Australia, and has been likened to a southern counterpart of the Gulf Stream of the Atlantic, must be directed against the west coast of New Zealand, tending to equalise the temperature in that region.

On the north-east coast of New Zealand, as far south as the

Bay of Plenty, there are further evidences of a current from the north to be found in the abundance of the flying-fish, the occasional visits of the true nautilus, and also of the argonaut, or paper-nautilus. Gigantic pods of a leguminous plant that grows in the Fijis are also frequently cast up in the same way that West Indian seeds are thrown on the coast of Scotland by the Gulf Stream. This current, however, although it reaches New Zealand, does not appear to pass down the east coast, as there is abundant proof of the existence of a steady drift from New Zealand eastwards towards the Chatham Islands. From this it appears that neither the west coast nor the north-east coast of our Islands are as favourable as regards temperature as the south-east coast of Otago.

A study of the kind of fishes which are associated together in different parts of the European seas, and a comparison of them with the fish found on our shores, will better enable us to form an idea of the probability of the ultimate success of the salmon experiment. Just as the land can be divided into divisions by the character of its productions, so the seas are marked by the distinct character of their marine fauna. The seas on the western coast of Europe, says Professor Forbes, in his "Natural History of the British Seas," may be divided into three provinces, two of which possess extremely distinct characters. The most northern is the Boreal Province, in which there are few species of fish, but great abundance of individuals, their pursuit affording the sole employment to large communities of fishermen. In this province the salmon skirts the coast-line and enters the rivers. Herrings abound in the surface-waters during their proper season; in 15 to 50 fathoms water the cod and the hake are plentiful; while from that depth to 100 fathoms, whiting, pollack, ling, and tusk employ thousands of men in most adventurous and perilous fisheries which are conducted in the deep sea far from land.

The most dissimilar province from the Boreal is the Lusitanian, which includes all the southern European seas, and is characterized by the profusion of sea-perches, like our kahawai (*Arripis salar*) and hapuku (*Oligorus gigas*). Sparoids, like the snapper (*Pagrus unicolor*) and scomberoid fishes, which include the mackerel, flying-fish, tunny, and John Dory, together with gurnards and mullets of various kinds, all of which are without representatives in the extreme north. On the other hand, we remark the absence from this marine province of the salmon and all the many species of the cod kind, which are plentiful in the former.

Between the Boreal and the Lusitanian the fish are of a mixed or intermediate character—a sort of neutral ground, into which the northern fish from the Boreal region are attracted during the winter season, while the southern genera are represented during the summer months. These are the seas round the British Islands. In them the turbot, flounder, sole, cod, haddock, and whiting reach the greatest perfection, while the herrings and the salmon amongst the northern forms, and the mackerel, sea-bream, and red-mullet from the south, combine to render the British seas the most prolific and profitable in the world.

If we now compare the assemblage of fishes which we find in the New Zealand seas with those in the European regions, we find that on the whole they represent the characteristic forms of the southern or Lusitanian Province, from which the salmonidæ are almost entirely absent; in other words, that our New Zealand fishes resemble those which are found on the coast between Madeira and the Bay of Biscay more than those which are caught about the north of Scotland. If we contrast the thirty odd sea-fishes that are fit to be used as food in New Zealand, we have amongst the constant residents on all parts of our coasts the hapuku, tarakihi (*Chilodactylus macropterus*), trevalli (*Caranx georgianus*), moki (*Latris ciliaris*), aua, rockcod (*Percis colias*), wrasse (*Labricthys bothrycosmus*), and patiki or flounder (*Rhombosolea monopus*), and while the snapper, mullet, and gurnard are only met with in the north, the trumpeter, butterfish, and red-cod are confined to the south. But, with the exception of the patiki and the red-cod, none of these are representatives of fishes that are common even in the south of Britain, while from the more northern seas, where the salmon are most plentiful, similar fishes are altogether absent.

We are now in a position to form some conjectures why hitherto the attempt to acclimatise the salmon has not succeeded. With regard to the whole of the west coast and part of the north and east coast, it is an attempt to introduce a fish which is rarely, if ever, found in the corresponding fish zone in European regions. The young of the salmon have been turned in thousands into many rivers of the South Island—rivers in every way most appropriate as a habitat (as shown by the rapid growth both of the salmon and brown trout) until they reach the smolt stage. It is after their departure from the rivers that their difficulties begin. At the period of the year when the young smolt, putting on his silver coat, makes his journey to the sea, the mouths of nearly all our rivers swarm with many of the sea-perch tribe, especially the voracious and active kahawai. Whatever happens to him, the fact remains that the smolt never returns. What becomes of him is unknown; but, out of the tens of thousands—I might say, hundreds of thousands—of young fry that have been turned out into the magnificent rivers of New Zealand, not one thoroughly authenticated instance can be, as yet, adduced of the smolt's return to the river, and its capture as a veritable salmon (*S. salar*) by the angler.

It is too early yet to speak with certainty of the success or failure of the experiment in the Aparima. The sea at its mouth is probably four or five degrees colder than at the mouth of any other river in the colony, with the probability of the absence, to a corresponding extent, of the sea-perch class of fishes. When we remember how few of many millions of the eggs of any fish ever come to maturity, or how few of those that do so reach the adult stage, it is premature to give a decided opinion on the subject. It is the wish and hope of every colonist that the experiment may succeed, and the men who have worked hard to make it so do not belong to the class that is easily discouraged. One actual experiment is worth a bushel of theories, and our only course is to wait and see what the future may have in store.

But there remains to account for the fact that the trout thrive so well, not only in the brackish water at the mouths of all the rivers running out on the east coast, but in the actual sea itself. That the common brown trout frequents not only the brackish water, but is found in harbours, many miles from the mouth of any river, is now established beyond doubt. The late Mr. Arthur, in a paper read by him before the Otago Philosophical Institute in 1883, alludes to the disappearance of the largest trout from the Water of Leith except during the spawning-season, and their evidently resorting to the salt water of Otago Harbour, since trout of the common *S. fario* species were being constantly caught in fishermen's nets in the Bay. The experience of all anglers is that the extreme mouths of the great snow-rivers, even as far as it is possible to throw the spinning-bait into the surf, are the surest places to find the largest and best-conditioned trout. In December, 1891, when the graving-dock in the harbour at Lyttelton was emptied, a trout weighing 15lb. was caught, this being several miles from the mouth of the nearest river. In March, 1892, Mr. Arthur Perry, of Timaru, a very successful angler of many years' experience, informed me that trout may be seen on almost any calm day playing round about the stones of the breakwater in the Timaru Harbour; yet there is no fresh water or river-mouth within miles; in fact, nothing but the open sea. The state of the tide, too, nearly always determines the time of the sport. As soon as the tide begins to turn the trout seem to come with it into the river. For two or three hours the fun will be fast and furious, till before high water is reached they cease to take. More will be said about this later on, in the description of the Rakaia and other snow-fed rivers.

It is evident, then, that the sea as well as the rivers is

suitable for the salmonidæ. The answer to the objection seems, however, simple. Trout of small size, even so small as 1lb. in weight, are seldom, if ever, caught in the salt water. They do not seem to affect the brackish estuaries and the mouths of the rivers, where kahawai and hapuku abound, until they are large enough to take care of themselves, whereas the smolt descends when a few inches in length, consequently falling a prey to the predatory fish. The difficulty, but an insuperable one, would be at an end if the smolt could only be induced to stay in the fresh water till it was big enough to take care of itself.

CHAPTER II.

Snow-rivers and Rain-rivers — Advantage of Division — Different Appearance and Size of Trout — Reasons for their Enormous Growth—Different Growth in Different Rivers —The Character of their Food-supply — Their Feeding Times—Capricious Character.

GENERALLY speaking the snow-fed river produces a trout both larger and different in appearance to that found in the rain-fed stream. Of course in the deep pools of any stream a large trout can nearly always be found, but the average size of those taken from rivers like the Rakaia and the Waitaki is greater than that of trout taken from the smaller rain-fed streams. In the small streams the trout still keeps the appearance of its English parent, being marked with many deep red spots, though these become less numerous as it increases in size, and at last disappear altogether. In the snow-rivers the absence of red spots, even in small trout of $\frac{1}{2}$lb. weight, is almost universal, their whole appearance being of a bright silver colour, those taken near the sea having a larger number of bluish-black X-shaped spots.

There is no doubt that the yearly growth of our trout is, as compared with the English fish, immensely greater : this arises from the supply of food, which, as will be shown hereafter, is enormous, and has had much influence on their habits. It is difficult, in the case of rivers that have been stocked for a number of years, to speak with certainty as to the exact yearly growth. The late Mr. Arthur, in a paper read before the Otago Philosophical Institute in 1878, calculated the yearly growth of certain fish procured by him as follows : Shag River, female trout, $16\frac{1}{4}$lb., yearly growth $2\frac{3}{4}$lb. ; Water of Leith, male, $12\frac{1}{4}$lb., yearly growth $1\frac{1}{2}$lb.; Lee Stream, 5lb., yearly growth 1lb. ; Deep Stream, 8lb., yearly growth $1\frac{1}{2}$lb. ; and Upper Taieri, female, 6lh. 6oz., yearly growth 1lb. These data were arrived at from a knowledge of the period which elapsed between the young fish being turned into the river and its capture.

Trout were first placed in the Omarama River, a tributary of the Ahuriri, in 1875. In December, 1879, it was first fished by Mr. J. A. Connell. During one day, two evenings, and one afternoon he caught fourteen trout, weighing 27½lb., or an average weight of 1·96lb., the heaviest fish 7lb. Mr. Arthur also fished the Omarama, and killed five trout, weighing 13¼lb., or an average of 2·7lb. each. This would give their greatest yearly growth at about 2lb.

The Waitaki River was fished by Mr. S. Thompson in 1879, when he caught twenty-two fish, weighing 23¼lb., or an average of 1·06lb. each. It was first stocked in 1869, and again in 1874. Taking Mr. Thompson's largest trout, which was 5¾lb., as one put in during the latter year, its yearly rate of growth would be 1·15lb.

Fulton's Creek is a small stream, the upper part of which descends rapidly from the mountains through bush, and its lower waters flow gently through the alluvial plain of the Taieri in deep, long reaches. Trout were first put into it in 1869, and none since then. In July, 1881, Mr. Deans caught two beautiful females, 18lb. each. The least possible annual growth of these two fish would therefore be about 1·5lb.

In March, 1883, Mr. McKinnon killed, with native minnow, in the Puerua a trout of 22lb. As trout were first turned into this river in 1873, the least possible yearly growth of this fish would be 2·2lb.

The yearly growth of lake trout is very great. Hayes Lake is very full of trout, some being supposed, from appearance, to weigh over 20lb. Mr. Arthur mentions one of 28lb. which was poached out of Butel's Creek, running into Lake Hayes, in 1882. Now, trout were first put into the streams round about the Wakatipu in 1874; so this very large trout must have grown yearly at the astonishing rate of 3½lb.

The Wakatipu Lake has a stock of trout from 2lb. to 25lb. The water is very clear, and they can be seen in shoals on the margin of the lake at Queenstown Bay, near the Town Creek, the Peninsula Reef, and other places. They are exceedingly fat and lazy, and, on account of the clearness of the water and abundance of food, are difficult to catch with the rod and line. The largest yet taken weighed 28lb. In November, 1880, a trout was caught at the head of the lake which weighed 16lb. 4oz., and, as this was probably one of those liberated in 1874, its yearly growth would be nearly 2¾lb.

Lake Heron, in the Ashburton County, holds an astounding number of large trout, which can with difficulty be caught with rod and line. Two were shot in 1884 weighing nearly

34lb. each. Many large trout have been taken by spearing them. The method used is to pull slowly in a boat round an arm of the lake after dark with a spear and a torch. In 1887 one man obtained in this manner 112 fish, weighing over 7cwt. Lake Heron is rather inaccessible, being distant from the Mount Somers Railway-station some thirty-four miles; and, finding the fish would not keep, he salted them down in barrels. Either the hot weather was too much for them or his method of curing insufficient, and they all went bad. To utilise them, however, with true colonial thriftiness, he gave them to his pigs, with the totally unexpected result of killing every pig —a thoroughly deserved reward for poachers and spearers.

Taking Mr. Arthur's table for some of the above rivers, and others, we find as follows for the growth of New Zealand trout between 1878 and 1883 :—

YEARLY GROWTH OF TROUT, 1878 TO 1883.

	Lb.		Lb.
Omarama	2·00	Otaria	0·75
Kakanui	1·00	Mimihau	1·00
Waikouaiti	1·07	Pomahaka	1·08
Fulton's Creek	1·50	Waitahuna	1·12
Lorsell's Creek	1·66	Manuherikia	1·00
Puerua	2·20	Hayes Lake	3·50
Waiwera	1·50	Wakatipu Lake	2·73
Waipahi	1·00		

If the growth of trout is abnormal, the reason is found in the equally abnormal food-supply as compared with English streams. Most of the rivers swarm with various kinds of fish-food, and in such abundance that it is a wonder the trout ever take the angler's bait. Smelts (*Retropinna richardsoni*), whitebait (*Galaxias attenuatus*), both anadromous, are in such countless millions at the mouths and for some distance up the rivers that frequently they are impaled on the triangles of a Devon minnow or a phantom when spinning. Crayfish, the young of the sea-mullet (*Agonostoma forsteri*), fresh-water molluscs (Limnœa), flies with their larvæ, beetles (especially two kinds, the well-known brown-beetle (*Odontria zealandica*), in appearance like a cockchafer, and the well-known little green-beetle), grasshoppers, bullheads (*Eleotris gobioides*), a fish in appearance somewhat similar to the loach, and minnows (*Galaxias fasciatus*), all these make such a feast for the hungry trout that it is no wonder it grows at a prodigious rate. Mr. Arthur speaks of the stomach of one trout that he opened which contained thirty-eight bullheads and smelts. In 1889 I was fishing with Mr. Willock, accountant to the Bank of New Zealand at Christchurch, near the mouth of the Rakaia River. He caught a fish weighing 11lb., and on opening its stomach we found no less than eighty-five bullheads and smelts, each of which could be easily identified ; besides these, there was a mass which was too far digested to be distinguishable as fish. I have often opened the stomach of fish and found them so distended with smelts and whitebait that the difficulty would be to find room for the live bait with which they were taken.

The migratory habits which our trout have adopted have no doubt been occasioned by their finding out that an enormous supply of food comes in from the sea in the spring and gradually ascends to some considerable distance up the rivers. In the early part of the season, in October and November, the fish are always found in best condition as near the mouths, of the snow-rivers especially, as it is possible to get with safety to the angler. More than one fatal accident has happened in an attempt by the over eager fisherman to go too far out into the surf. Two years ago, in the spring of 1890, two anglers were fishing from a boat at the mouth of the Waitaki River, and both were drowned through the boat being carried by the current into the breakers. In October, 1891, Mr. Pilbrow, of Ashburton, was fishing with a party on the south side of the Rakaia River, close to its mouth. The trout were evidently well on the feed, and, finding the nearer he got to the sea the better the sport, he took up a position at the extreme end of a shingle-spit on the sea-shore. A strong south-west wind was blowing behind him : a larger wave than usual suddenly passed over the spit, and with it he disappeared ; he was carried out to sea and drowned. This unfortunate occurrence was entirely the result of inexperience. To the angler who is ordinarily prudent there is no danger : if he will only remember that the great snow-rivers of New Zealand are treacherous in the extreme; that what is a shallow ford to-day may be a deep channel to-morrow; that he never ought to attempt to cross any stream with a swift current where it is deeper than up to his knees, and not even then if the stones are large or the shingle is loose; that it is better never to attempt to cross at any time when the water is thick unless accompanied by a companion on whose experience he can rely ; that under no circumstances should he attempt to cross a rapid shingle stream at night; and, lastly, that it is better to lose even a 20-pounder than his life—he can obtain sport to his heart's content, and with a certainty of its not ending in such a catastrophe as that which occurred in the season of 1891-92 at the mouth of the Rakaia.

Experience shows us that our New Zealand trout has all the idiosyncracies of its English progenitor, with those which it has acquired in New Zealand added, and these last are by no means few; in fact, English experience, except of course the acquired skill of handling a fish after hooking it, is not worth much. Dry fly-fishing in most rivers, to use a colonial phrase, "isn't in it" with the sunk fly or grasshopper or cricket, and it is almost impossible to say what time of day the trout will come on the take. Mr. Arthur mentions that in Otago, during the day, the middle portion in particular, when the sun was brightest and everything apparently against the angler, he found by far the most trout come to his lines; and, moreover, that he found them shy in the morning and evening; but in this particular his experience is contrary to that of every angler with whom I have discussed the subject. It is true that in the very early part of the season, perhaps throughout October, when the morning and evening are raw and cold, the trout will take only for an hour or two in the middle of the day, during the warmest and brightest portion; but after October, the universal experience of anglers is that the early morning and late evening are far better than any other time; in fact, from about noon till 3 or 4 o'clock it is mostly labour in vain. Here, as in England, of course, much always depends on the day and the state of the water. In Canterbury, given a south-west wind, however cold, the trout will be on the feed, more or less, throughout the whole day; but with a north-easter, the worst of all winds, and, unfortunately, the most prevalent, the trout will not move in the middle of the day. Where the river is affected by the tide, other considerations come in, about which more hereafter; but, speaking generally of rivers unaffected by tide, if you can get nothing before 12 o'clock you may reckon on a poor day's sport, unless you fish late in the evening. Sudden changes in the weather and wind, for which New Zealand is justly famous, will often bring the trout on the feed for a little while, and then the angler should make the most of it, for it will certainly not last. I have fished with two or three friends before now for hours without stirring a trout, though we must have passed over hundreds; suddenly the fit seized them, and, although fishing some distance apart, for an hour or so we all caught fish; then the blight came again, and not a fin would stir. A slight change of wind, or perhaps a difference in barometric pressure, was the only way one could account for it.

The most extraordinary case of capricious behaviour in trout occurred to a friend and myself when camped on the Waitaki, just south of the railway-bridge, on the great south line. This part of the river is particularly well stocked with trout, and anglers, from Oamaru especially, have made very many large bags between the railway-bridge and the sea, for which see the description of the Waitaki, further on. We started about 5 o'clock in the evening, and each took a fish of about 6lb. within the first quarter of an hour. This augured well for future sport; yet we fished this river for the following four whole days without having a single fish even run at the baits we used. Spinning-tackle of all kinds, live bait, creepers, worms, every size and colour of fly was tried without the least effect; the water was in as good condition as we could wish it; the wind, if not favourable, was not adverse. All to no purpose. Miles of this river were fished, and on the last day of the four, besides ourselves, eight other rods took up our parable. At the end of the day not one fish was there to the ten rods, and we had to leave at last, satisfied, at least, that we had probably made a record which no angler will desire to break.

When, however, really on the take our trout, big and little, seem to abandon all caution, and come boldly at a fly or bait of any kind if not very clumsily presented to them. Even where much fished they are far less shy than the Home species, especially the large fish; and this arises, probably, from their having practically no enemies in their habitat after they attain a certain size. Their worst enemy, the black shag, gormandising brute as it is, and the greatest destroyer that small trout have to contend with, has little power over a fish that has once attained to a weight of a couple of pounds, and we have seen, under favourable conditions, this would be at the end of its second year. Sometimes a large trout will dash at the spinning-bait two or three times, and even in some instances get hooked (though this is exceptional) yet come again at it, and eventually be caught. In fly-fishing, except on windy or cloudy days, the middle portion of the day is the worst possible time. On the other hand, after the beginning of November, it is almost impossible to begin too early in the morning or fish too late. Trout-fishing is allowed during the whole twenty-four hours, as during the greater part of the summer the large trout are seldom caught in the day time except in rapid water. The last hour before dark and the hour immediately preceding the dawn are favourite times for the trout to feed: then there will be a lull, and about 10 o'clock in the morning they will often feed for an hour, even on the worst days. In February, 1890, I was stopping on the banks of the Clarence—a somewhat late river—which was at that time clear and low. The banks of the Clarence are very high, and large trout could be seen motionless at the bottom. Regularly, however, between 10 and 11 o'clock every morning they would leave their haunts and cruise round for food. This would last, perhaps, for an hour, and then there was no further movement till 4 or 5 o'clock in the afternoon. Fishing, too, with the creeper or grasshopper I have found the same thing—early morning and late evening, but nothing done in the middle of the day after the first month of the season.

The largest trout yet caught with rod and line was taken by Mr. Lambie, of Leeston, fishing in Hall's Creek, near Leeston, with live bait, and weighed $26\frac{1}{2}$lb. Another very fine fish, weighing 25lb. was taken from the same place by Mr. Beetham, the Resident Magistrate for Christchurch, the latter being far the more perfect in shape and general proportions. In the Selwyn many fish have been caught up to 18lb.

The rapid growth has also its effect on the shape and general appearance of New Zealand trout. This will be seen by reference to the accompanying plates.

CHAPTER III.

Licenses—Free Trout-fishing—Fishing-tackle—Rods—Lines—Casts—Flies—Spinning-baits—Crickets—Grasshoppers—Creepers, &c.—Winds—Effect of Electricity or Electric Light—Rules and Regulations for Trout-fishing—Camping-out.

THE first thing the angler has to do before commencing his season's sport is to obtain a license from the acclimatisation society for the district in which he happens to be. This costs £1, and gives him the right to fish in any part of the colony for the whole season merely by getting it indorsed by the secretary of any other district. The regulations are mainly the same throughout the colony. A conference on this subject was held at Oamaru in March, 1892, when it was recommended that there should be one set of regulations for the whole of the South Island, and probably these will be drafted and come into force shortly. The most important of the regulations are : The holder of a license can fish in any waters from the 1st October till the following 15th April, using as baits either the natural or artificial fly, artificial spinning-baits of any kind, any small indigenous fish, grasshoppers, spiders, caterpillars, creepers, and worms, but not mussels or shellfish of any description. Every bait must be used in its natural state. No medicated bait, or bait prepared with any chemical, may be used. No fish whatever of the salmon kind may be taken except trout. Only one rod and line may be used at the same time. Trout under 8in. in length must be returned. The holder must produce his license when asked to do so by any Ranger or person holding and showing a license, and, if required, exhibit to any such person the contents of his creel or bag and the bait he is using. None of the salmonidæ may be bought or sold except as provided by the regulations for taking lake-trout. The penalty for a breach of the regulations or for selling trout is not less than £1 nor more than £50, and a license will be refused to any person implicated in the illegal sale of trout or in breaches of the regulations. The full text of the regulations is given in the appendix.

Throughout New Zealand trout-fishing is practically free.

The river-beds belong to the Crown, and, provided the angler does no damage to fences, he can go and camp in them where he pleases. This applies more especially to the great snow-fed streams. For some of the smaller rain-rivers leave is necessary, but it is seldom, if ever, refused. In addition to the river-bed proper, a reserve a chain wide has been made all along the banks of many rivers, on both sides.

Two rods are sufficient, the one for fly, the other for spinning and live-bait fishing. Every angler has his own ideas as to the best material for rods, and, generally speaking, if the rod is suitable, it does not matter much of what wood it is made, but if the angler does not object to some little expense, split-cane rods with steel centres are certainly preferable to any other kind. Not that a split-cane rod is essential, but with our large trout it is certainly more lasting and keeps its shape better than any other. Constantly having to hold strong trout, running from 1lb. to 5lb., will try any fly-rod, and the steel centre not only gives the necessary extra stiffness required, but keeps the rod from losing its straightness and elasticity. On a favourable day when spinning it is common to get one or two 10lb. or 12lb. fish; and with a heavy current and a strong fish, unless the rod is fairly stiff, most of the best part of the day, and probably several good pools, are spoilt in playing fish which a little extra stiffness in the rod would have saved. Common English fly-rods are of very little use; they were never made to hold New Zealand trout. Luckily, the makers at Home know now what is required, and rods can be obtained in any of the large towns in the colony suitable for the trout they have to deal with. Messrs. Hardy Brothers, of Alnwick, make a special New Zealand rod.

A 12ft. fly-rod is the most convenient length, and, if slightly stiffer than the common English fly-rod, will hold any fish that is likely to take a fly. The largest I have ever taken in fly-fishing weighed 6½lb. Mr. Irvine, fishing with me some years ago in the Hororata, took a trout of nearly 8lb., and Mr. St. Barbe, in the upper waters of the Opihi, a very fine female weighing 8lb. Both these were caught with a fly. Dr. de Lautour, of Oamaru, informs me that Mr. Flint, fishing in the Ahuriri, a tributary of the Waitaki, in 1888, took trout with a fly from 3lb. to 10lb., but these cases are very exceptional, and, as a rule, trout taken with a fly seldom exceed 5lb. Many of our smaller streams are bordered with New Zealand flax, the well-known *Phormium tenax*, and it requires good tackle and a strong rod to keep the fish from getting underneath the roots. For open water a 9ft. cast is not too much, and either two or three flies; but for narrow streams, where the banks are fringed with flax or any kind of bush, from 4ft. to 6ft. of gut is ample, and two flies at the most: for the latter kind of stream it is better not to use more than one. The size and strength of the gut depends more on the nature of the stream than anything else. It is quite useless fishing with very fine tackle where it is necessary to hold steadily on to your fish. On the other hand, of course, in open water, the finer the tackle, consistent with a reasonable degree of strength, the better. In streams much overgrown with flax (the flower of which seems specially invented to promote profanity) the best way to succeed in killing heavy trout is with a good stiff rod and strong line and a yard only of gut: let the fly or beetle sink well in the rapid current, and keep the top of the rod moving with the bait. Strike immediately you see the bait stop, and hold on as much as you dare, or Master Three-pounder will be under the bank and in the flax-roots in a twinkling, his subsequent proceedings interesting you no more.

The angler is strongly recommended to buy his flies in the colony, as the tackle-makers are now well acquainted with what is required, and can tell the flies most suitable for the different streams. A rather larger fly, with stronger gut than the usual English trout-fly, is necessary with most rivers, though a great variety is not essential. The best for general use are the March brown—especially the Irish March brown, with claret hackle, in the early part of the season—the Moorfowl, the Black Ant, the Whirling Dun, the Red Spinner, the Coch-y-bondu, a medium-sized Green Drake, the Brown Moth, and, above all, the Yellow-tailed and Red-tailed Governors. In November and December, when the common brown-beetle (*Odontria zealandica*) is about, any fly something of the pattern of the soldier-palmer, with a thickish brown body, is very deadly: the beetle itself, too, placed on the hook of an artificial fly and well sunk, will often account for some of the best fish when simply flinging the fly in the most artistic manner on the surface is of no avail. Thirty to forty yards of almost any kind of trout-line is enough.

Cicada-fishing in Otago and Southland in January, February, and March gives first-rate sport. A medium-sized hook should be used. Put the hook through the cicada and try its point on your finger before using it. Throw it like an ordinary fly, keeping ready to strike the moment you see or feel it taken; then look out for squalls, as the biggest fish take it greedily. With many of the smaller streams with confined banks, more especially those covered with flax, the water goes for some distance underneath your feet, and it is there in the day-time that the big trout lie. They cannot see the bait if only trailed on the surface of the part of the stream visible; but sink it and they will come out from underneath after it.

The angler will find out in a day's fishing of this sort that very light tackle is of little use.

Fly-fishing, where the river is open, is, however, far the best sport; one can use a fine cast and two or three flies. In rivers like the Upper and Lower Selwyn, in the streams about Masterton, and in many of the rivers of the South the angler will often take fish with the fly up to 3lb., 4lb., or even 5lb., and all his skill is tested with these with light tackle. No prettier sport can be found than this, and baskets of 15lb. to 30lb. are common.

By the sides of the rain-rivers, underneath the stones (if these are large enough), close to the water's edge, in the early months of the season, will be found a black hideous-looking grub. This, called the creeper, or black-fellow, is the larvæ either of the dragon-fly or of a large fly with wings folded close to its back when at rest—in fact, identical with the English stone fly, only three times the size. Sometimes only one, sometimes two or three are found under the same stone. Turn over a few stones, pick up the grubs, and put them into a little tin match-box. They are quite harmless, and, ugly as they look, they are the finest possible lure for trout. Use a hook about No. 6 in size, and pass it through the little square black head, letting the grub hang down its full length. About two or three small shot, or, better still, 2in or 3in. of fine lead wire wrapped round the gut, should be placed about 8in. or 10in. from the hook. Throw, if possible, up stream into all the little eddies, at the sides of the most rapid waters, and if the trout are on the feed at all you will soon find it out. These grubs are very tough, and can be thrown like a fly without any fear of their coming off the hook. Strike directly you see the line stop or feel the slightest tug, or the trout will have eschewed both the bait and the hook. In any rapid water these grubs are taken freely, but it is little use fishing with them in the deep pools.

In the decayed logs of the black-pine another larger grub can be found, but generally a small axe or tomahawk is necessary to obtain them. These, too, make a famous bait; they are larger than the water-creeper, but equally efficacious, and should be used in the same way, only with more care, as they are somewhat softer.

The largest bags of fish are made either by spinning- or live-bait fishing in the great snow-rivers. Some marvellous takes in the season of 1891–92 will be found in the accounts of the Rakaia and Rangitata Rivers. For fishing in all these rivers a 14ft. or even 15ft. rod is required, and at least 60 yards of line, with a good check reel. Anglers are rather given to increase unnecessarily the length of their lines; and wonderful accounts are told of how a mighty trout ran out line up to 150 yards and then got off, and no wonder. Let any one measure out 150, or even 100, yards of line, and imagine a trout at the end of it in a good strong snow-river current, and see what chance he would have of exercising any control over its movements. In ten years I have never had any fish run out the whole of my line, and I never use more than 60 yards. Of course one must follow the fish; stand still and you will soon find much more line than this gone; but, unless you are casting from the end of a shingle-spit, between two streams, it is nearly always possible to follow with perhaps a little wading. Within reasonable limits, the nearer you keep to your fish the better. Let it once get beyond 60 yards from the end of your rod and your chance of killing it is very small. Once, fishing in the Waimakariri, in Canterbury, I had to cross the river to avoid losing a very strong female fish. Below where it was hooked was a rather high bluff, and two willow-trees jutting out into the stream. The fish made off down stream, taking an occasional rest in mid-water, and then turning again down stream till it gradually came nearer and nearer to the willows. The other side of the river was clear, but to cross it was evidently necessary to go in over the tops of my waders. Another rush, and there was nothing for it but to cross or lose the fish. It is very disagreeable going in over the tops of one's waders, but it was that or nothing, and the fish appeared much bigger than it eventually turned out to be. Across, therefore, it was, at the top of a big rapid, getting wet up to my waist. Luckily the fish halted for a minute or two while I was in mid-stream, and then started again down the river; by that time, however, I was ashore, and eventually succeeded in landing it about 150 yards below the crossing. It turned out to be a very handsome female fish weighing $12\frac{1}{2}$lb. That is the only time when I felt I should like to have had more than 60 yards of line; it was the stiffness of the rod saved me. With a weak rod, fish, line, minnow, and trace would all have gone.

For the snow-rivers, almost any spinning-bait, from 1in. to 2in. or $2\frac{1}{2}$in. long, will succeed, including the archer and other tackle for spinning dead baits. A spinner much used in the colony is known as the whitebait phantom, and is an imitation of the little silvery anadromous fishes that yearly ascend the rivers: sizes 0 to 2 are the best. The Devon is a very killing bait, either silver or brown. Geen's spiral minnow, too, the angler should certainly have: sizes 2in. and $2\frac{1}{2}$in. Last year on some occasions trout would take it readily when they would not look at either the Devon or the whitebait phantom. The 2in. size is the best, and the colour either gold or silver, or silver- and-brown. With one gold-coloured Geen I killed numerous fish in the season of 1891–92, and at last lost it in a very large one in the Rangitata.

The mounting of all the spinning-baits is most important. The tackle-makers at Home will persist in sending them out with the triangles mounted on single gut. This is perfectly useless. It is a mistake to buy anything that has not the hooks mounted on strong double gut at least, and even treble-twisted gut is not too heavy for the triangles. There is not the least necessity for this mounting to make them clumsy if properly done. Strong single gut, finished off with pretty floss silk, looks *so* nice; but it is about as much use for a day's fishing in a snow-river, when the fish are on the feed, as single horsehair: one good fish will put your spinner *hors de combat*, and you have all the bother of either sitting down to put on fresh triangles just at the time when the trout are taking fast, or constantly putting on a new spinner. Medium-sized gimp, the colour of your minnow, will do, but it is more clumsy than treble gut. Triangles of a medium size are better than large ones, but they should be fastened with silver wire. What is called bouquet wire answers well, with the coils not too close together round the shank of the triangle or it will rust. It can be fastened off in the ordinary way by taking two or three turns round the shank and pulling it through. To the spinner attach about 1ft. or 18in. of either strong salmon-gut or double-twisted gut, ending in a loop.

For a trace use 6ft. of strong salmon-gut or double-twisted gut. A trace of treble gut is too clumsy, except for night-work. The fine steel casts are admirable, but throwing the minnow overhand they are apt to get kinked if not very carefully used. In underhand throwing, however, they very seldom kink, and certainly show less in the water than any other kind of trace. To the end of the trace nearest the minnow should be attached a box-swivel. This allows one to change the spinner without removing the trace, and is a great advantage. Of course, in the centre of the trace another fixed swivel should be placed.

To spin well requires considerable practice. It can be done with great success both in the snow- and rain-rivers, especially in the early months of the season. In the great snow-streams —like the Rakaia and Waitaki—the trout often lie close alongside the bank, and are often caught within a foot of the shingle-edge. When they lie like this little art is required in throwing, and spinning then is far easier than when the trout are deep down in the centre of the stream or far out in some back-water or stream you wish to reach. Throw up stream if you can, letting the bait come down well and slowly across the stream. The great mistake beginners make is to spin too fast: as long as you keep the bait twisting that is all that is necessary. Of course it is impossible always to throw up stream; the current may be too strong or the position unsuitable. Often, however, when walking up-stream the great back of a trout may be seen in the early morning close in shore; throw your bait ahead of him, and as it comes past him he will be almost certain to go for it; but, if you walk up-stream and try and fish down to him, he will either see you or the line or something, and be scared and gone long before it reaches him.

Anything unusual in the way of a spinning-bait seems to succeed, and if the angler has some novelty it is always worth giving it a trial. In 1885 Mr. Beetham, the Resident Magistrate at Christchurch, fishing in the Lower Selwyn, using a 4in. Devon, had marvellous success during that whole season. Not being able to get any at the tackle-shops, I had some specially made, at considerable expense. The next season I did nothing with it, nor has either he or I been able to do any good with it since. This is one of those incomprehensible problems that are constantly cropping up in trout-fishing without ever being solved. Why should the trout of one year take a certain kind of bait freely and the next refuse to be attracted by it at all; yet in this case it certainly was so. Both of us gave it many a trial the next season, with the poorest success, and then, putting on a 2in. Devon or whitebait phantom, would take trout after trout. In the season of 1891-92 nothing seemed to succeed so well when the trout were off the feed as Geens' spiral minnow, yet whether it will do so next season is a problem of the future.

In spinning, too, success depends on varying the depth. On some days I have found it far better to keep the bait close to the surface, more especially with the whitebait phantom. At other times the trout would be lying very low, and much more lead is then required to sink the line, so as to spin as close as possible to the bottom. In the snow-rivers there is little danger of going too low. The water nearly always shows the proximity of snags, which can thus be avoided, and the greater part of the bottoms of the shingle-rivers consisting of small boulders and sand, from which the minnow is easily released if it gets caught.

Of all methods of taking trout with the rod and line fishing with live bait in still water is the easiest and least skilful, and yet it is one of the most effective. For this the beginner has only to obtain a supply of bait, a rod and line, and a 6ft. or 9ft. stout gut cast with a No. 6 hook. The bait used is either the bullhead (*Eleotris gobioides*), or the New Zealand minnow (*Galaxias attenuatus*), or smelt (*Retropinna richardsoni*), the Maori inanga, and these can generally be obtained with a hand-net in the river itself or in the small creeks close by. Pass the hook through the extremities of both lips of the bullhead or inanga, beginning with the underlip; then throw it out to swim about in a deep pool, and abide results. This is the favourite method of the night-fisherman. In the evening he will be seen beginning his day's sport as the sun goes down, his rod in his right hand, a bag, gaff, and lantern on his back, and a bait-can in his left hand, taking up a position at some favourite deep pool, and probably staying there far into the night or even till the dawn, for the hour before the dawn is one of the most taking times for the monster trout. Here the novice and the veteran meet on equal terms, the only skill required being to keep the bait from staying too long on the bottom of the river, as inattention to this particular results in the immediate capture of the unwelcome eel.

It is gruesome work this night-fishing; yet if the live-bait fisherman only sticks to it long enough, on a favourable night (and what is or is not a favourable night no one can tell till next morning), he seldom returns home without a good fish or two, and not rarely is rewarded with a heavy bag, for the trout of 10lb. and upwards feed seldom at any other time in water where there is no current. The last three miles of the Selwyn River, in Canterbury, are very deep, and with scarcely any current, just enough to keep the bait moving, and this is the favourite haunt of the all-night angler. Up till last year, probably not less than a ton weight of trout was annually taken out of these last three miles of water, the supply of fish being constantly kept up from Lake Ellesmere. Last season was a failure only on account of drought and the river getting choked with weeds, but these have now been carried away by flood, and probably in the season of 1892-93 this river will be as prolific as ever.

One can never say at what particular hour the trout will feed by night, the last hour of day and the first before the dawn being generally found the best, but at any moment the feeeding-time may begin, often without any apparent reason. Frequently on a summer night as many as twenty or thirty rods will be engaged on this particular stretch of river for hours without one of them getting a run; yet hundreds of fish must have seen the bullhead or inanga, and refused it. Three or four hours of this will choke off most of the fishermen; but an enthusiast or two, with true Sassenach persistency, will fish on in hopes of something which will gladden their hearts. Sometimes it comes, and then two or three great trout will be taken in an hour, weighing, more, perhaps, than he can conveniently carry; or that something wanted may not come at all—hungry, and weary, and sleepless, and fishless, he sees the dawn breaking over the everlasting hills, and, anathematizing the wretch who invented all-night fishing, vows that he has been made a fool of over it for the last time.

One piece of advice may, however, save the reader many a profitless hour: electrical disturbance seems to put the fish down at once; the first flash of lightning, and, whatever time it is, give it up and go home; it is quite useless to keep on; for that night the sport is over. Mr. Knubley, of Timaru, informs me of a peculiar instance of the effect of any sudden light on fish, which was shown when the "Ringarooma" warship was in Timaru exhibiting her electric search-lights. On the Opihi the fish were taking fairly well on that evening up to about 9.30, when the ship began to throw her search-lights on the clouds at an angle of about 45°. This was sufficient to cast a reflection of the flashes on the Opihi River, a distance of about eleven or twelve miles, when the fish at once left off taking. The same effect was noticed by several anglers, when the "Ringarooma" was in Lyttelton Harbour, with regard to the trout in the Selwyn River.

Fishing by day with the live bait in rapid water is a different affair, as much judgment is required both in throwing the bait and making it travel at the proper depth over the most likely parts of the river, carefully guiding it past the corners of snags, a sure hiding-place of trout, and generally making it present as natural an appearance as possible. With the enormous number of bullheads and other small fish in our streams it is a matter of wonder that the trout ever take at all, unless it is that the sight of a little fish in apparent difficulty is too enticing. On a good, rough, windy day they often, however, feed freely. The first throw should be made at the head of the rapid, just above where the water begins to break; then, immediately having executed the throw, lower the point of the rod to within a foot or two of the water (the mistake beginners invariably make, after they have made the cast, is to hold the rod up as though fishing with a fly). Walk slowly down the rapid, keeping opposite the bait as it is carried down the stream, letting it sink without check, as, if taken at all, it will be when it is nearly close to the bottom. Do not strike if you see it stop; it may be only a snag; and a few seconds will not make matters worse. It may be a fine trout, and if you strike immediately, instead of giving the fish just a moment to gorge it, you may, if it is a bad day, lose your only chance. Then gather in any slack line and strike; you will find yourself stuck fast in one or the other. A single hook passed through both lips of the bait is, as a rule, to be preferred to two placed one above the other, as in what is known as "Stewart tackle." It catches far less in obstructions. Of course, if you are using two hooks you must strike immediately the bait is seized. Occasionally, however, the fish will come short. By this is meant that they will, instead of taking the bait by the head or shoulders, as they always do when they mean to swallow it, take it by the tail or middle of the body. Sometimes several trout, one after the other, will catch hold of it, run a short distance, and then drop it, evidently trying to bring the angler into disrepute. If you find them doing this whip on a second hook about an inch above the first, putting the bait, of course, on the upper hook, and strike immediately the bait is touched.

A small bag, to carry one's impedimenta—tackle-book, luncheon, &c.—is preferable to a large creel, unless a man or boy is employed to carry it. Where the trout run small a creel that will hold the probable catch is useful; but when spinning, if any fish are caught, they will probably average 6lb. or 7lb., and even a couple of fish of this size are very burdensome to bear on one's back all day. The best way, if returning over the same ground, is not to attempt to carry the fish, but to hide them in the tussock, or cover them up with shingle, and then collect them on the return homewards. The worst accident that can happen to you is to start fishing three or four miles from your camp in the early morning, and begin in the very first pool with an eight-pounder: as you are not returning over the ground again, you have then to carry this burden, with probably sundry others, all the time you are spinning. Add to this your bag, gaff, bait-can, and perhaps a mackintosh, and you will find a few hours of it quite enough. It will be seen, then, that it is better to begin with the water nearest your camp or hotel, if it is all equally good, and fish back to it.

When spinning a gaff should always be used, as the triangles get sadly entangled in the meshes of the landing-net, and then one kick of a big fish will tear them off, even if on the strongest gut; but for fly-fishing a landing-net is preferable. Indispensable, too, is a disgorger. Take a foot of ordinary fencing-wire and beat one end flat, file a notch in this end, and turn the other end into a bow, and you have the most perfect disgorger you can wish. The ordinary shop ones are not long enough if the trout is hooked far down in the throat, and it is inadvisable to put your finger into the mouth of a ten-pounder. *Experto crede.* For convenience of reference, a list of the angler's equipment will be found in the Appendix.

Waders and good heavy boots may be looked upon almost as a necessity, certainly for all shingle-streams, and are better brought or obtained from England. Anderson's No. 1 quality for heavy wear are admirable. Good quality waders can be bought, it is true, in every large town; but waders, to last, should fit the foot, or they soon leak, and become useless. The boots or brogues must have plenty of large square-headed nails, well clinched before the sole is put on to the uppers, otherwise the heavy walking on the shingle drags them out of the sole. Of course, waders can be dispensed with if one does not mind being wet all day, but this means certain rheumatism sooner or later. Gum-boots are used commonly, and answer very well, for night-work, but in the day their black colour attracts the heat and makes them almost unbearable.

Camping out in New Zealand in the summer months is most enjoyable. Very little rain, comparatively speaking, falls between November and March—certainly nothing that an ordinary calico tent and good fly will not keep out. Of course, from time to time there will be a wet day or two, but seldom any continuity of wet weather. Everywhere wood and water can be found in abundance; and, above all, one of the greatest charms of tent-life is the entire absence of snakes or harmful insects of any kind or description. The only poisonous insect known is the katipo, but these are so rare that in ten years I have never seen one. The tired angler may throw himself down in the heaviest tussock or flax without the least fear of harm, or may sleep at night in his tent in absolute security. What a boon this is they alone know who have lived in a country where venomous reptiles are common; but the fact remains that, while snakes and centipedes are very common in Australia and Tasmania, not one single specimen can be found in the whole of New Zealand. With a horse and cart and camp outfit the angler can penetrate into gorges containing some of the finest scenery in the world, or can camp in the river-beds, and have no long journey to make after his day's sport is over.

A tent made of calico is quite sufficient, and better than canvas, on account of its less weight; one 8ft. by 10ft. is ample for two or even three persons. The fly should always be at least 2ft. longer each way than the tent, as it not only keeps out the rain better, but provides a shelter for rods, spare tops, and other impedimenta which are likely to get damaged, or which take up useful room inside. It is better to have the tent-pegs of ½in. iron and about 1ft. long. The tent should always be well guyed with at least four guys, and pitched so as to be well sheltered from the winds, more especially from the north-west and south-west, the north-west in Canterbury being always the one to guard against. The pegs can be carried in a small bag to prevent their rusting the tent, and rolled up with it. For poles, if bamboo cannot be obtained, two ash sticks, such as are used for hay-fork handles, from 6ft. 6in. to 7ft. long, will be found most convenient, being both strong and light, with a ridgepole of the same material. A ridgepole may, however, be dispensed with altogether by having a rope sewn along the top of the tent, passed over or through the uprights at each end, and fastened to strong guys. Where the camp is intended to be permanent for some months a ridgepole is perhaps necessary, together with another separate one for the fly; but for ordinary fishing-expeditions, in the summer months, this is not required and means adding very much to one's impedimenta without any corresponding advantage. A general list of requirements for a camp are given in the Appendix. The angler can select from this what he chooses to take, and what he thinks luxurious and can be done without. Though it may be considered a luxury, a camp-bed, with pillow attached, 80in. long by 36in. wide, which can be obtained from Silver and Co., in Cornhill, at a moderate price, is, unquestionably, an immense boon. These are made of indiarubber, and, by means of a pair of bellows (which are made for and given with the bed), can be inflated in two or three minutes. The whole thing, when the air is exhausted, takes up less space than a blanket, and relieves the angler, tired with his day's sport, of the necessity of cutting fern or brush, and at the same time gives all the comfort of the most luxurious couch. The season for trout-fishing lasts from the 1st October to the 15th April, but the best camping-months are from December till the end of February.

THE NORTH ISLAND.

General Observations on the North Island—The Auckland Province—Rainbow Trout—Tauranga—Hawke's Bay—Taranaki, or New Plymouth—The Wellington Province.

IF the coasts of New Zealand were not visited by cold currents and the climate cooled by the extent of seaboard, there is no doubt that a great part of the North Island certainly would be too tropical for the well-being of the finer varieties of the salmonidæ.

As it is, their introduction into the South Island has been a marked success, most of the rivers in that Island being well enough stocked to afford magnificent fishing for trout. In the North Island they have been successfully introduced into most of the rivers as far north as Taranaki on the West Coast and the inland portions of Hawke's Bay on its eastern side.

As the work of acclimatisation progresses year by year, rivers further north are being stocked, and no doubt the central system of rivers running north will be stocked at high altitudes where found suitable.

With the exception, however, of the Wellington Province, no part of the North Island affords as yet the sport which can be obtained in the South, in the Provinces of Canterbury, Otago, and Southland. The reasons are not far to seek. The climate of the South is far more suitable for the majority of the salmonoids; they were introduced at a much earlier date; and in the South, the Maoris, both from paucity of numbers, and from their rights having been well defined or extinguished as regards the rivers many years ago, interfere but little with them. In the North Island, on the other hand, the majority of the Europeans live either in or in the immediate vicinity of the towns. While the Maoris in the South Island are only about 2,000, there are probably 40,000 north of Wellington. Much of the land, too, is still held by them, and this interferes sadly with the work of acclimatisation. Nevertheless, many thousands of fry have been distributed, and it is only a question of a few years before abundance of sport will be attainable. In some few places it can be got even now, and we proceed to give the angler some information about these.

The acclimatisation of fish up to the present time in the Auckland District has not been so successful as in the South, principally through the earlier efforts of the society having been directed to the introduction of salmon and the English brown trout, which, as experience has long proved, are not quite suitable for its climate. Of late years the Auckland society has turned its attention to the introduction of the Californian rainbow trout (*Salmo irideus*), which is capable of withstanding a higher temperature than the English species. It has been successfully established in several localities, and has increased sufficiently to justify the society in throwing the streams open for fishing, which was done for the first time in 1891–92. The following are the chief localities :—

1. Lake Takapuna, within a few miles of Auckland, and easily accessible by steamer and coach. Rainbow trout, 9lb. in weight, have been taken in the lake.

2. Henderson's Stream, about thirteen miles from Auckland by rail.

3. The Upper Thames River, with its tributaries—the Oraka, the Waimakariri (which must not be confounded with its famous Canterbury namesake), and the Mangawhara; also, in the same district, the Pokaiwhenua, a tributary of the Upper Waikato. The chief liberations of rainbow trout have been made in these streams, and they will probably afford excellent fishing in a year or two. Several anglers had good sport this last season, landing fish varying from $\frac{1}{2}$lb. to 4lb. Good hotel-accommodation can be obtained at Okoroire, Oxford, and Lichfield, near some of these streams, and the district is easily reached by rail from Auckland.

4. The Upper Waipa and its tributaries, near to Otorohanga. Trout are said to be plentiful here, but the district has not yet been fished. The above-mentioned localities are the principal ones, but rainbow trout have been placed in very many other streams, although too recently to admit of successful fishing.

The Tauranga District.

The Tauranga society was formed in 1881, and already has turned out many thousands of trout into the principal rivers in the Bay of Plenty, and also the lakes in the Rotorua and Taupo districts. Already the Rivers Waiau, Wainui, Waipapa, Te Puna, and Te Pere-a-tukahio have been stocked, together with a number of rivers and creeks which flow into Lakes Rotorua, Tarawera, Rotoiti, Taupo, and several other smaller ones. From the small number of European residents the society has as yet been unable to determine with certainty what success has been attained.

The Hawke's Bay District.

In this district the rivers have been stocked for many years, but no fish have been caught by rod and line till the season of 1891-92. The rivers are the Manawatu and Ngaruroro. The Manawatu is available by rail from Napier to Danevirke—about four hours' journey. It rises in the Hawke's Bay ranges, and, though affected by snow to a certain extent in the early part of the season, is practically a rain-river. It has an open shingle-bed. The creeper can be found under the stones by the water's edge, but fish have hitherto been taken with the spinning-minnow, from 2lb. to 10lb. The grasshopper can also be obtained, and is a taking bait. There are good hotels at Danevirke, within two miles of the river—a river easily waded when in proper fishing-order, and is good all through the season. Leave should be asked for, but there is no difficulty in obtaining it. A southerly wind is the best.

The Ngararoro lies to the west of Napier, and can be approached by coach to Kuripapanga, where there is a good hotel. It is a snow-river, with plenty of fish, and is a shingle-stream. Fish have been taken with minnow and grasshopper from 2lb. to 6lb. It is affected by snow up till about Christmas. The green-beetle is found to be a killing bait.

The Waipapa and Tukituki have both been stocked with trout, and fishing can be obtained in their upper waters. They are open river-beds, only slightly affected by snow in the early part of the season. To the Waipawa by train from Napier, about two hours. For the Tukituki to Waipukurau, about two hours' journey, five miles past Waipawa. Leave should be asked, but there is no difficulty in getting it.

Taranaki, or New Plymouth District.

Fair sport can be obtained in several rivers in this district. Possessing as it does innumerable streams, that radiate from the base of Mount Egmont, this district offers many opportunities for the introduction of the Salmo family. These streams are somewhat short, and pursue their courses seawards with pools and rapids in quick succession, thus providing the true home of the acclimatised trout.

If the angler takes up his head-quarters at Hawera, a township on the main line between Wellington and New Plymouth, there are numerous streams within his reach. He can either take the early train to Inglewood, in the neighbourhood of which there are some half a dozen streams, all containing a few trout, or can drive a few miles along the main road. Going in the latter direction, the first principal river is the Waingangoro, running about four miles distant. This takes its rise direct from Mount Egmont, and is a fine trout-stream; it is full of pools and shallows, quick runs, and stickles, but though many trout have been turned into it, they have not shown up so well as might have been expected: some good fish, however, are reported near Eltham. Three miles further on the road runs the Inaha, a pretty little stream, with rather a preponderance of deep water. The fish in this stream are not very plentiful, but a few have been taken up to 3lb. or 4lb. in weight. Half a mile further on the angler will cross the Te Apuni, differing in character considerably from either of its predecessors, being much clearer and brighter. This is fairly well stocked, especially some ten miles up from the sea, where good baskets are occasionally made. The fish are clean and bright, and very pink. One fish has been taken from this river weighing 15lb.

Another mile and the Waiokura is reached. This is hardly large enough to afford really good fishing, but it deserves notice on account of its being the first stream in the district in which trout showed up in any quantity. A short time back it was remarkably well stocked, and was used as a reserve for the distribution of fish to other streams, but in 1892, during the dry weather, it became very low, and much illegal work was done. Close to this is Manaia, a small township on the Waimate Plains, where good accommodation can be obtained, and most of the streams above mentioned are within easy walking-distance.

If the railway is chosen the first stopping-place should be Stratford, some twenty miles from Hawera. The Patea flows through Stratford, and is a mountain-fed stream. It is as yet only fairly stocked; but, as it has a small breeding-establishment on one of its tributaries, it will not be long before it should be well supplied with trout. Thirteen miles further on is Inglewood, the centre of the principal fishing in the district. Six well-stocked rivers are within fair walking-distance, between Inglewood and Stratford.

The first three of these are the Ngatoros, being within a stone's throw of each other, and all fairly well stocked with fish up to 6lb. or 7lb. These streams all take their rise from

Mount Egmont. Large baskets are continually caught in these rivers, and they are the favourites with anglers at present. Three miles from the township these unite into one large river, which in its turn flows into the Wanganui, which contains large trout.

The Piakau is a little smaller than either of the Ngatoros; but, if anything, is better stocked. Trout have been taken up to 6lb., and are remarkably game and strong. The Makataura, a mile further on, is rather a faster-running stream, and has a very good reputation. Lastly, Waitepuka, though little frequented on account of its distance, should be mentioned, as its trout are remarkably strong and bright.

THE WELLINGTON DISTRICT.

The sport in the Wellington Province, more especially in the Wairarapa and Hutt Counties, is far superior to anything else in the North Island, and excellent baskets can be obtained over nearly a hundred miles of running-water in Wairarapa, and upwards of forty miles in the Hutt County.

The active manner in which the culture of trout was taken up by the Wellington and Wairarapa Acclimatisation Society in 1884-85 has done much to render the trout-fishing in these districts equal to that to be obtained in the earlier-stocked rivers in the South, and opens up a wide field for the enthusiastic angler, who, by starting here early in the season and following it up in the late snow-rivers of the South Island, can have the cream of the fishing during almost the whole open time.

To any one able to afford the time and small expense of travelling, in most parts of the colony this opens up a perfectly feasible way of enjoying some of the finest trout-fishing to be had in the world, either small fish taking fly in the little streams, or grand fish in the larger rivers, at times caught with the fly, but more generally with the minnow and grasshopper or cicada in midsummer. Generally speaking, the best baskets are made early and late in the season in the North Island, while the colder snow-fed rivers in the South are later, and afford capital sport in midsummer, when the fish in them are in strong condition.

Turning to the fishing in the Wairarapa and the Hutt districts, almost every variety of stream the most fastidious angler can wish for flows out of the Tararua Range of mountains, the backbone of the southern end of the North Island. Before, however, treating of the streams separately, perhaps a few hints as to the rods and tackle most suited to these waters will be of use.

Fishing-tackle.—The fishing can be divided into two classes—light fly-fishing in small streams, for trout from 9in. long to

4lb. in weight, and fishing in heavier water for larger fish. As in the other parts of the colony, a 12ft. greenheart or split-cane rod, with a good easy reel, and 40 yards of light silk acme, or any other trout-line, will be found convenient, a landing-net, and a stock of good casts and flies. Good gut is the main thing, as it does not keep well in the dry climate of parts of New Zealand, and wants careful attention. The flies that seem to take best in the smaller streams in Wellington are the red-tailed governor, March brown (quill body preferable), and red-quill gnat, black and red hackles early and late in the season; and in midsummer, the stone-fly, alder, black gnat, blue upright, and others of this kind. As a rule, a cast made up with a March brown, governor, and red-quill gnat is as good a trial as can be made in a stream. Towards the middle of summer, when the water is clear, and lots of cicadæ and grasshoppers are about, fish of all sizes will be most readily taken with them when they will not look at an artificial fly.

Turning to the fishing in the larger rivers, an outfit consist of a 14ft. 6in. greenheart rod, good reel, and 60 yards of good silk or dressed line will be found useful. A supply of single grilse-gut traces (which must be of first-class quality), and one or two heavier traces for evening fishing will be necessary. For minnows, soleskin and whitebait phantom Nos. 0 to 2, whitebait phantom in silk, same numbers, a few small quill minnows for fine water, and one or two green phantoms will be found sufficient for trolling. The principal food of the trout in most of the larger rivers is the inanga, or silvery, of which there are several varieties. This fish is imitated by the phantom whitebaits, which are, as a rule, the most deadly minnows to spin with. In all the rivers is the bullhead or "bully" (*Eleotris gobioides*), on which the trout feed largely, and this is imitated by the brown and dark-blue phantoms, or Devon. As a rule, trolling with these natural fish on a flight of hooks is far more deadly than using the artificial imitation, especially in clear water, but it involves the usual bait-cans, and other trouble. In addition to the trolling appliances mentioned above, it will be well for the angler to have a stock of flies, casts, and some dapping-hooks (Nos. 6 or 7), so as to be able to turn his attention to any lure the fish are seen feeding on. Most of the anglers here prefer to troll with a limp fly-rod, and cast a light minnow overhand, using it to fish far off and lightly in clear water.

Almost all the rivers in the Wellington District are free, the property-owners very seldom objecting to any man fishing, provided he shuts gates, and does no damage to fences.

Starting from Wellington as a base of operations, the first trout-stream of any magnitude that is reached is

The Hutt River.

Rising in the almost-impenetrable fastnesses of the Tararua Range, this river flows through many miles of rocky gorges practically inacessible to the fisherman till it emerges into a fine valley, where the river affords capital fishing for about fourteen miles, till it enters the Wellington Harbour. Large fish are known to exist in almost every pool among the mountains; hence the stock in the lower reaches is being continually replenished by fish dropping down and running up from the sea, the trout in this river being extremely migratory in their habits.

Starting from Wellington, eight miles by train brings the angler to the Lower Hutt Station, about half a mile from the lower waters of the river before it enters the harbour. Here the water is much fished, but there is always a chance of getting a heavy sea-run fish, especially when the water is coloured with flood. The fish in this part of the river are, however, shy, and hardly so numerous as higher up the river.

Following up from the Lower Hutt good fishing can be had all the way up the valley, in which there are five stations on the railway-line—Belmont, Hayward's, Silverstream, Wallaceville, and Upper Hutt—and fair accommodation can be had on the opposite side of the river to the railway, at the Taita, between Belmont and Hayward's, to which place a coach runs to meet nearly all the trains at the Lower Hutt. Higher up the river there are hotels at Wallaceville and the Upper Hutt, close to some of the finest reaches of the river.

Following up the railway-line to Kaitoke Station, above the Upper Hutt, a beautiful branch of the Hutt River—the Pakuratahi—can be reached by about two miles walk, and there are several farmhouses on the banks where accommodation can be had. This is one of the finest fly-streams about the district, full of excellent trout of good quality and very game. A trip to this stream is a capital outing, and very enjoyable, especially to the angler fond of fine scenery. There are between three and four miles of good fishable water available before the stream enters the rocky gorges of the mountains.

Before going over the mountain-range which separates the Hutt from the Wairarapa, there are several small streams where capital trout-fishing can be had, notably,

The Wainui-o-mata Stream.

This stream is situated about twenty miles from Wellington, and from it the water-supply of the city is drawn. This was the first stream stocked in the North Island, and has steadily maintained its character as a good fly-fisher's water, from which phenomenal baskets have often been obtained. It is perhaps the most certain water to go to near Wellington, and it is not often that the careful angler returns from it with an empty basket. The trout run up to about 4lb. weight, but are rather muddy at times in flavour, and not equal in quality to those taken from the streams that rise in the higher mountains.

To reach this favourite resort, take train to the Lower Hutt, get a horse or trap at the stables, and drive over to Wainui, a distance of about ten miles. Accommodation can be arranged for with the settlers there; but it would be better for the angler paying his first visit either to get some one who knows the river to go over with him, or get a man from the stables to drive him over, show him the ground, and arrange about accommodation, and bring him back.

There are two streams on the West Coast running into the pretty inlet at Porirua where fly-fishing can be had, but, as they flow through small enclosures and private grounds, it is necessary to ask leave of the landowners before fishing. The Porirua Stream can be reached from Porirua or Tawa Flat Station, and the Horokiwi by coach from Paremata Station (six miles). They are both small streams, in which fine tackle is necessary.

For all-round fishing there is no place so good in the North Island as the Wairarapa district, through which a large number of fine rivers and streams flow for a distance of some thirty to forty miles into a shallow lake before entering the sea. As an all-round centre from which most varied fishing can be obtained, Masterton, a town of about three thousand inhabitants, and the site of the central fish- and game-rearing establishment of the Wellington Acclimatisation Society, is the best. From here a number of rivers can be reached by easy trips. The Ruamahanga is the largest, and full of big fish, with capital runs and pools most suitable for minnow early and late in the season, but affording good fun at times with fly, and excellent sport in midsummer with locust or grasshopper.

The Waipoua and its branches are excellent water for fly; and the Waingawa, a fine, rapid river running into the Ruamahanga from the high ranges, is full of large, powerful trout.

There is no lack of river on all sides, and the best thing the angler can do is to go to the Club Hotel and consult the genial host, and the curator at the fish-ponds, Mr. Ayson, who will soon put him on the track of good fishing.

Lower down the Wairarapa Valley there are two fine rivers— the Waiohine, near Greytown, and the Tauherenikau, at Fernside, near Featherston, both of which are full of large, powerful trout, more given to fish than insect diet. The angler cannot complain of any scarcity of good fishing, and if once settled at Masterton, with a trap at his disposal, can fish a new part of some river every day for months.

There is a stream at Eketahuna, the present terminus of the Wellington railway-line—the Makakahi—in which fishing was opened in the season of 1891-92, and a good many fish of from 4lb. to 11lb. taken.

Before closing this brief sketch of fishing in the North Island it might be useful to point out that, as a rule, the trout feed badly in south-easterly weather, and that light northerly winds are preferable for sport. Here, as elsewhere, the best sport is had when there is a little colour in the streams, after rain, especially while the water is rising and not too high.

Some of the takes of the season 1890-91 in the Wellington District, are as follows:—

Name.	Date.	Number of Fish.	Weight.	Name of River.
	1890.		Lb.	
Mr. J. Eman Smith	Sept. 15	35	45	Wainuiomata.
Mr. C. P. Skerrett	"	12	15	"
Mr. F. D. Dyer	"	15	17	"
Mr. Smith	Sept. 16	15	16	"
Mr. Dyer	"	25	33	"
Mr. Skerrett	"	10	13	"
Mr. Smith	Oct. 15	20	27	"
Mr. Dyer	"	16	18	"
Mr. Skerrett	"	13	15	"
Mr. Smith	Nov. 3	15	17	"
Mr. Dyer	"	18	21	"
Mr. Skerrett	"	16	19	"
Mr. Smith	Nov. 9	34	43	"
Mr. Dyer	"	25	32	"
Mr. Skerrett	"	20	27	"
	1691.			
Mr. Smith	Feb. 1	20	28	"
Mr. Dyer	"	17	20	"
Mr. Skerrett	"	12	15	"
Mr. Smith	Dec. 7	10	35	River Hutt.

Twenty-four days' fishing for three rods in the Wainuiomata alone gave 548 fish.

THE SOUTH ISLAND.

THE NELSON DISTRICT.

There is good sport, both with fly and minnow, to be obtained in the Nelson District, the chief streams being the Wangamoa, Wairoa, Takaka and Aorere, Motueka, Maitai, Happy Valley, Riwaka, and Lake Rotoiti.

Wangamoa.

The Wangamoa River, to its nearest fishing-point, is eighteen miles from Nelson, and to Oliver's Half-way Accommodation-house, twenty-three miles. The direction taken from Nelson is through Wakapuaka and Happy Valley and over the Wangamoa Saddle.

The tourist can either go by coach three times a week or in his own conveyance. This river is fishable its entire length of about ten miles. It is composed of rough rock and gravel, with a good fall to the sea, and runs through the forest its entire length, except where here and there there may be a clearing. It is a capital fishing-stream, being broad and of walking-depth.

The largest take of fish yet recorded is thirty-five, weighing on an average $1\frac{1}{4}$lb. All these were taken with a fly. The largest weighed 6lb., but there are fish here and there weighing 10lb. All the fish taken with the fly are fair-sized fish. It is best adapted for the fly, but the minnow can also be used. An ordinary morning's take should be about a dozen. All along the bank of the river are capital camping-places.

Wairoa.

The Wairoa River, situated in the County of Waimea, is to the nearest point twelve miles from Nelson. There are two hotels on its banks—the Bridge Hotel and the Brightwater Hotel. The proprietor of the Bridge Hotel is an angler, and can give all the desired information requisite. This river is a rain-river and is about forty miles in length. The lower part, being about a third of its length, runs through level country with farms on each side, and its bed is composed of gravel and silt. The upper part is bounded by bluffs on each side, on the top of which is the forest, and its bed is composed of huge rocks, boulders, and gravel. Only when it is low, and it is that as a rule in the fishing-season, is it crossable. It is a broad, swift, and deep river. Trout abound in numbers, and many of them are large, the smallest usually caught being 3lb. in weight, and ranging up to 20lb. The average fish caught run up to 8lb. The only baits which are used are the live bait and spinning-minnow. The river, from its mouth to its source, is public property. The takes from this river are not large in number, the greatest number taken being five, which weighed 30lb. The fish taken from this river approach very nearly the colour and flavour of the salmon. There is a large tributary of this river—the Roding—which is of the same character to a very great extent, and from which some very large fish have been taken. In the course of a year or so these two rivers will be renowned for their fish. To get access to the Roding River it is necessary to cross the Hope Saddle and drop into the Aniseed Valley. There are no hotels here, but, as a public road runs along its bank, excellent camping-grounds are available. These two rivers have always been most attractive to the Nelson angler, as well as others.

Takaka-Aorere.

The Takaka and Aorere Rivers are each respectively fifty and seventy miles by sea from Nelson, and several small steamers run there during the week.

The Takaka Valley is magnificent, and is renowned for its limestone caves. There are several hotels here where the tourist can put up. The Takaka River has only lately been stocked with trout, but in the course of a few years should make a first-rate trout-stream. On the lower part of the river a boat can be used, and many places of interest can be seen. In parts there is the usual wild look, offering a strong contrast to others, where the farms reach to the river's edge.

The mouth of the Aorere River is the Port of Collingwood, where there are two hotels. This river rises in a very mountainous district, and has a rapid fall through the native forests. The bed is composed of large rough boulders, but can be forded at convenient places. The river is a succession of whirls and eddies, with an occasional deep hole. Many large trout have been seen here, but, as it is only within the last ten years that they have been liberated, much cannot be expected at present. No angler has yet paid it a visit, and as it is practically undisturbed, the trout should multiply and increase.

The surrounding district has always been a place of interest, on account of its gold-mines. Here it is that the famous limestone caverns are seen to perfection. Eight miles up the river is an exceedingly small township, with one hotel, named Slateford. From this point, without reference to angling, the tourist should find scope for investigation.

Motueka.

The Motueka is another fine river, with a gravel bottom, and bids fair to be a magnificent trout-river. It has a course of about forty miles, and to the nearest station on its banks is about thirty-two miles. There are two ways of getting to this river—firstly, by taking train from Nelson to Belgrove, and thence by coach over Spooner's Range, down Norris's Gully, and to the Motueka Bridge. Bromell's accommodation-house is on one side of the river and Robertson's on the other. Mr. Robertson being a successful angler, could give very important information to anglers. The river contains only very large trout, varying from 6lb. to 14lb. The spinning-minnow and live bait are the only kinds of bait used. The other way is by Belgrove, and then by coach up the Waiiti Valley, over Reay's Saddle, and to Stewart's accommodation-house, which stands on the banks of the Upper Motueka waters. A tributary, the Motupiko River, is gained by travelling from Stewart's to David Kerr's station. It is here, in the head-waters of that river, that the trout are so plentiful. They vary from ½lb. to 1lb., and take the fly with great eagerness. Permission would have to be asked to pitch a camp here, but no objection would be raised; on the contrary, a hearty welcome would be given. Both rivers have a gravelly bottom, and the fish are very numerous and in good order. At the head-waters of both these rivers there is magnificent mountain scenery. As the bed of the River Motueka is Crown land, no difficulty arises as to fishing. A careful angler should be able to land three or four large trout from the Motueka and at least two dozen from the Motupiko in a morning's fishing. To be successful in the Motueka River the wind should blow up stream.

Lake Rotoiti.

Lake Rotoiti is situated about sixty miles from Nelson, and about four miles from Tophouse Telegraph-station. Take train from Nelson to Belgrove, drive thence through Waiiti Valley, over Reay's Saddle, thence to Stewart's accommodation-house, over Motueka River to David Kerr's station, thence to Tophouse. This lake is seven miles long by two miles wide. The River Buller runs into and out of it. Mr. John Kerr's station is in close proximity, from whom all particulars can be obtained, or from the proprietor of the Tophouse accommodation-house, who has a large boat on the lake. The lake is full of very large trout, the smallest caught or seen weighing 10lb., whilst the largest are, probably, from appearance, more than 25lb. These large fish are caught at the mouth of the Buller, running into the lake, or at the top end of same river, running out of the lake. The live bait or a frog is the only bait which they will take. There are also the Canadian whitefish, which Mr. John Kerr liberated in the lake, and these are seen in great numbers from time to time, but few if any have been taken so far. To the angler this spot is one full of interest, irrespective of angling. This fine sheet of water lies at the foot of the St. Arnaud Range, the steep precipitous sides of which here and there, covered in dense forest, stand magnificently reflected in the smooth lake. In the whole district is magnificent scenery, and to the alpine climber it would suggest something more than ordinarily new. The best plan is to camp on the margin of the lake, or, in the alternative, board at the Tophouse accommodation-house. The Buller is partly a rain-river, and partly a snow-river. The upper part of the Buller is strictly called the Rotoiti River.

Maitai.

The Maitai takes its source about eleven miles from the City of Nelson, and has two head-waters, called the Maitai North and South, each meeting about two miles from their source, and forming one of the prettiest rivers for trout-fishing. This river is entirely a rain-river, with a moderate fall to the sea. The bed of the river is formed of large and small gravel, with light vegetation on its banks, and is crossable at any point. At present the river is not fully stocked with trout, but in 1891 a large number of fry were released, so that in a very short time it should be in first-rate fishing-order. In 1890 and previous to that date large takes of fish were made. One angler alone took thirty-two fish, weighing on an average ¾lb. Lately the largest fish taken weighed 6lb., but this is very occasional. The artificial fly is the favourite bait used, as it is the most deadly; but many anglers use the spinning-minnow. The river is open to the public.

There are many hotels in Nelson from which the angler can make his starting-point, without the necessity of camping out. This river should be a favourite resort for tourists, as it contains many spots of beauty, the New Zealand ferns and forests growing in perfection.

Happy Valley.

The Happy Valley River is ten miles from Nelson, and is about nine miles in length. The banks are open for a considerable distance, but about five miles of its course is through scrub and bush, which, however, is not so bad but that the angler can fish it. This river is also a rain-river with a gradual fall to the sea. The bed is composed of rock and gravel, and can be crossed at nearly any point. The trout thrive well, and many excellent takes have been made. An ordinary angler should be able to land at least ten fair-sized trout. The fish are not large, but all are in good condition. They vary in weight up to 4lb. An angler has obtained as many as thirty-eight fish, weighing 25lb. The fly is used by all anglers, with few exceptions, and these use the spinning-minnow.

There are no hotels on the banks of this river, but there are excellent camping-places from the middle waters upwards.

The surrounding country is pretty, with abundance of ferns and native bush. To the angler it is an ideal place, with its profuseness of vegetation, and seated in the heart of the mountain-land.

Riwaka.

The Riwaka River is four miles from Motueka, which latter place is sixteen miles by sea and thirty-two miles overland from Nelson. Take train to Richmond, thence by coach over the Montere Hills, through Upper and Lower Montere, and then to Motueka.

The angler can put up at the Riwaka Inn, which has all the usual conveniences.

The Riwaka River is about twelve miles in length, with a gravelly and shingly bottom. The river is fordable nearly at any point, with a succession of deep holes. There are some large trout, the biggest weighing 10lb. and from that down to 1lb. This place has had only an occasional visit from the angler, and although a few weighing about 5lb. have been taken with the fly, yet nothing can really be said of this river. It is known that numbers have been taken by illegal means, and it is said that at places there are plenty of fish.

The surrounding district is prettily laid out in small farms, and the farmers here would take great interest in giving any information, as well as making a stay perfectly agreeable.

North Canterbury.

Strictly speaking, the Clarence River is in the Marlborough Province, but has been made the boundary of the North Canterbury Acclimatisation Society's district, since it is more approachable and more easily stocked from the Canterbury side. It flows out of Lake Tennyson. The traveller should take train from Christchurch to Culverden, which is as far as the railway extends, then by buggy or coach to the hot springs on the Hanmer Plains. At the foot of Jollie's Pass, two miles from the springs, is a good hotel. From the hotel to the Clarence, over Jollie's Pass, is about five miles, the river being immediately at the foot of the pass. Here trout may be caught up to 10lb.; in fact, this is one of the best parts of the river for trout. Use either the fly or the minnow—the minnow for choice. The river is rather a late one. In the summer myriads of grasshoppers can be found on its banks—the sparrow and lark not yet having found their way over the pass—and these the trout take freely. As there is no hotel nearer than Jollie's Pass Hotel, a tent should be taken if it is intended to stay more than one day at the foot of the pass.

Following the river down for ten miles the angler comes to an accommodation-house. This is close to the river, and the fishing here is fairly good, both for fly and minnow. To make this trip comfortably from Christchurch a week or ten days are required, and an inquiry should be made as to the state of the river, as it is affected by snow. The trip, however, can be made very enjoyable by combining shooting with fishing, there being a supply of rabbits around the hot springs, and all down the Clarence Valley, which the proprietors are only too glad to have shot. The Clarence is one of the prettiest streams in New Zealand, having confined banks, with long rapids and deep pools. Just below the accommodation-house the Acheron flows into the Clarence. This is a beautiful mountain-stream, full of rapids and deep pools, but is not yet heavily stocked. From Jollie's Pass to its mouth, on the East Coast, the Clarence flows for nearly eighty miles, partly through wild and rocky gorges difficult of approach. It is crossed by a bridge about twenty miles above Kaikoura; but no trout have yet been taken out of this part of the river.

On the way to the Hanmer Springs, the Waiau lies close to the road, on the right. This is a large snow-fed river, and spinning only or live-bait fishing will succeed. This river was stocked many years ago, and the trout in it are now becoming numerous. The township of Waiau is distant only a few miles from the Culverden Railway-station by coach, and trout up to 7lb. have been caught there. Mr. Bayliffe, the Postmaster at Waiau, in 1891-92, took in the course of the season several fish, both from the Waiau and the Dog Creek, on the Lyndon Estate, which joins the Waiau River about a mile above the township bridge, the largest being $8\frac{3}{4}$lb.

The Mason, which joins the Waiau just above the bridge, is also well stocked, but in summer becomes very shallow, and, consequently, a prey to poachers.

Where the River Staunton joins the Waiau, about six miles nearer the sea, and also where the Leader flows into it, at Mount Parnassus, opposite the Cheviot Hills, twenty-five miles from Waiau, no doubt excellent sport could be obtained, but the angler must take a tent, and see that the river is in good order before he starts.

The Hurunui, distant from Christchurch fifty-two miles, is crossed by the northern line of railway, and is another snow-fed stream. Many trout have been turned into it by the Canterbury Acclimatisation Society, and probably good fishing could be obtained anywhere near its mouth. The mouth is, however, so far from the line of railway and from good roads —in fact, so little accessible in comparison with other streams in Canterbury—that it will probably be neglected for many years to come.

The Waipara, the Ashley, and the Kowai, all crossed by the northern railway, are of no value to the angler.

The Okuku is a small mountain-stream running into the Grey River, and thence into the Ashley. It is distant from Christchurch about thirty-seven miles—twenty miles by train to Rangiora or Ashley Stations, on the northern line from Christchurch, and thence by buggy. Rangiora is the best station to stop at, as a conveyance can always be obtained at one or other of the livery-stables there. The river was first stocked with trout about 1887, and has already yielded very handsome fish up to nearly 8lb. The river-bed is small, and confined by very high banks. The months of November and December are the best for this river, and the creepers which can be found under the stones close to the water's edge are the best bait. Later on the fly is taken freely, or a small Devon. For five miles above the White Rock Homestead the river-bed is open shingle, but then the gorge is reached, and continues for about six or eight miles. Trout were first placed in a well-known pool at the lower end of the gorge (known as the Picnic Pool), and have made their way through it into the more open water beyond. All through this gorge there is a succession of deep rapids and pools. By the side of the Picnic Pool, which is surrounded by beautiful bush, is a most convenient camping-ground; but leave should be obtained from the owner of the White Rock Station. From this point the fishing is good both up and down the river for about six miles each way. A 12ft. rod is the only one required for this river.

The Waimakariri is one of the great snow-fed rivers rising in the mountains beyond the Bealey, and flowing into the sea only a few miles north of Christchurch. It is well stocked with large trout throughout its course, a distance of some sixty miles. The Bealey, where there is a good hotel, is on the main road between Christchurch and the west coast of the South Island. By train from Christchurch to Springfield, thence by coach to Castle Hill and the Cass, and thence to the Bealey, is about eighty miles. The fishing near the Bealey is best about four miles from the hotel on the Christchurch side, but is not worth the long journey from Christchurch for it specially, since so many other streams nearer at hand afford better sport. The scenery, however, is superb, both here and at the Cass, and fishing excursions can be made either from the Bealey Hotel or the Cass Hotel; or the angler can take a tent and cart and camp practically wherever he pleases without being interfered with. Spinning with some artificial minnow is the least troublesome and the best method for these parts of the river, and trout have been taken up to 18lb. in weight. For many miles the Waimakariri runs through deep and almost impassable gorges of exquisite beauty, but impracticable to the angler, till it emerges on the Canterbury Plains. About three miles from Sheffield is a very fine bridge, known as the Gorge Bridge. To reach it, take the train to Sheffield; twice a week the train will run from Sheffield on to the bridge, on its way to Oxford. Just below the bridge, on the Sheffield side, is a small cutting, used as a blacksmith's shop years ago when the bridge was being built. Now it is overgrown with grass and tussock, and makes a famous place for camping. As the train goes within a few yards of the actual camping-place, the angler needs no cart to transport his luggage. Both below and above the bridge first-rate sport can be had with any spinning-bait, but before starting from town inquiry should always be made as to the state of the river, it being very much affected by melting snows.

Some miles below the Gorge Bridge the river-bed becomes

very wide, and the river breaks up into numerous streams. Most of these contain trout, but are not at present conveniently approached. During the lower part of its course the river runs chiefly in one large stream, from time to time breaking up into two or three smaller ones, these again uniting and again dividing in the manner common to all the great shingle-rivers. The nearest point of the river to Christchurch is about seven miles, a coach running the whole way several times a day. Here, at Belfast, on the North Road, is a good hotel; or the angler can camp on the river-bed. For some miles, both above and below the hotel, good sport can be got either with spinner or live bait, the fish running very large. In the last fortnight of the season of 1889–90 I took six fish from this part of the stream which averaged very nearly 12½lb., the largest being 14lb. and the smallest 11½lb. Two miles further on is White's Bridge, where there is another hotel and an abundance of trout; and a mile further on is Stewart's Gully, where the railway-line crosses the river. Very many large trout have been taken at both these places in the last two years with live bait and spinning. The current of the last few miles of this river-course is very slight, and it is the only one of the great snow-fed rivers that does not carry its shingle to the sea. From these causes, and from difficulty of access, the mouth, which in the case of the Rakaia, Rangitata, and the Waitaki is looked upon as one of the very best portions of the whole stream, is comparatively useless to the angler.

The Cust is another stream in the northern district, of quite a different character, being in reality a small brook of not much importance, but it merits mention through its affording good fly-fishing in the early part of the season. In the early days of trout-fishing in Canterbury, till the large rivers put it in the shade, it was a favourite with many. Even now, 15lb. or 20lb. weight of fish is not an uncommon take. It is almost entirely a fly-river, the March brown with claret body, the moorfowl, red spinner, and yellow- and red-tailed governors being the best flies to use. Take the morning train from Christchurch to Rangiora, thence by branch line to Moeraki or Cust. If Moeraki is chosen, fish up stream till Cust is reached; if the angler starts from Cust he can either fish down the river or walk for a mile up stream, and will then find two or three miles of good water full of trout up to about 3lb. Most of the Cust is private property, but the owners will always give leave. It is much covered with flax, and two flies, on a 6ft. fairly strong cast, are sufficient. The brown-beetle, well sunk, will be found a very killing bait in this little river in November and December.

The Avon, at Christchurch, is one of the very few rivers with banks almost entirely in private hands. It affords, above Christchurch, very good sport with the fly, but, except in the upper part of the Christchurch Domain and along Park Terrace, leave must be obtained. Its sources are only about four miles from the city, and, on account of its being so heavily fished, very large numbers of fry are turned out by the Acclimatisation Society every year. The trout do not run large, about 2lb. being the limit, with occasionally a much heavier one from some deep pool. Below the city sport is worthless. Very many perch have of late years been turned into the lower reaches, but, owing to the enormous shoals of sea-mullet—locally known as herrings,—which come up as far as Madras Street bridge, and devour the perch-spawn deposited on the weeds, they have not been much of a success.

The Selwyn is the favourite river of the Christchurch anglers. It is not affected to any appreciable extent by snow, and may be looked upon as essentially a rain-river. Its length is not greater than about fifty miles, and it flows into Lake Ellesmere, a large shallow lake of nearly seventy square miles. The fly-fishing in its upper waters and the live-bait fishing in its lower waters are some of the best in New Zealand. To get to its upper waters the train should be taken from Christchurch to either Glentunnel or South Malvern, at each of which places the traveller will find a good hotel. The river is close to either of these hotels. In the early months the fish will take either the creeper, found under the stones close to the water's edge, or the fly. There is an enormous stock of small fish, and till within the last year or two trout were commonly taken up to 5lb.; but the drought in Canterbury in 1889 and 1890 caused the river to be very low, and most of the large fish were taken out by illegal means. The large stock in the stream will soon, however, make it again one of the best fly-rivers in Canterbury. The angler should start at either Glentunnel or Whitecliffs and fish up or down, but it is of little use fishing more than two miles below Coalgate, as in dry seasons the river disappears into the shingle. With regard to its lower course, after a disappearance of nearly twenty miles it reappears below the Selwyn Station, on the main south line, but it is only after one or two wet seasons any trout can be obtained here. The way to get to the best part of the Lower Selwyn is to go from Christchurch to Ellesmere Station, on the Southbridge railway-line. The river is close to the station, and a train in the evening brings one back to Christchurch. There is no hotel near, but about half a mile down the river there is an accommodation-house where the angler, if he prefers to stop, can generally get food and lodging. The river here is suitable either for fly, spinning, or live bait. Springston, one station short of Ellesmere, is another convenient stopping-place for the Lower Selwyn. There is an hotel close to the station, where the angler can procure a conveyance to the Lake Ellesmere flat, a distance of five miles, and, as there is no nearer hotel than Springston, it is advisable to take a tent or to make friends with the Christchurch anglers, several of whom have built fishing-boxes near the river, on a piece of land that will probably be set apart for angling purposes by the Government. The fishing on the lake-flat for nearly three miles is quite free, there being a reserve a chain wide on each side of the river as far as its mouth. The water is deep, with a slight current, and for years has yielded an astounding number of trout of the very finest description. In each of the seasons of 1888, 1889, and 1890 not less than one ton of trout could have been taken out of this water. In 1891 a much less number was obtained, owing to the drought and the river getting choked with weeds. Trout of 10lb. and 12lb. were, and are, quite common in this water, but the fish take nothing but live bait and spinning. Trout up to about 6lb., however, can be caught in the rapids a mile or two above with a fly. On the 7th November, 1889, I made a basket of fourteen fish on this river, weighing 71lb. These were all caught in the day-time, but many are taken by fishing late at night, some enthusiasts keeping on all night till dawn, and generally being rewarded by a good fish or two sooner or later. Any one who has tried the all-night system, however, and had two or three successive blank nights, will not envy them their success. In 1888, a party of friends and myself built a fishing-box, and our example has been imitated by several others, till now the lake-flat presents quite a village-settlement-like appearance, and much advantage to the railway and neighbourhood results therefrom. As the warm weather comes on in the spring the fish come up the river from Lake Ellesmere, so that a constant supply takes the place of those captured, and accounts for the enormous numbers taken. A 5lb. or 6lb. trout is looked upon as a small one. The largest I have ever seen taken weighed 18lb., but this was an uncommon size, the more common sizes being from 8lb. to 12lb.

Hall's Creek is a small stream also running into Lake Ellesmere, the nearest station to it being Leeston, on the Southbridge line. It possesses the distinction of having produced the two largest fish ever taken by rod-line—one by Mr. Lamby weighing 26½lb.; the other by Mr. Beetham, the Resident Magistrate for Christchurch, weighing 25lb. The last was the more perfect fish in every way. Only night-fishing, with live bait, seems to succeed in this little stream.

The Rakaia is one of the largest of the snow-fed rivers coming from the ranges across the Canterbury Plains, its riverbed of shingle being in many places more than two miles across. In the *Field* newspaper of the 9th January, 1892, this river was described as "one of the pretty streams which run across the Canterbury Plains." Its chief feature is ugliness; its only merit being that from the Gorge at the foot of the mountains to the sea—a distance of forty or fifty miles—it is stocked with the finest trout, seldom caught under 2lb., and not uncommonly attaining a weight of 12lb., 15lb., and even 20lb. In the spring the north-west winds, which are very hot, melt the snows in the great ranges, and it is then not uncommon to see the whole river-bed filled with water from bank to bank. But ordinarily the river is confined to three or four large streams, breaking up constantly into numerous smaller ones, always shifting their position; the main stream may be here to-day, but to-morrow a heavy spate comes down, and when the river recedes to its ordinary height it has shifted a mile from its former bed.

Such is the Rakaia—constantly changing its course, carrying annually many thousands of tons of shingle to the sea. It nevertheless is the home of some of the finest trout to be found in the world. It is difficult to say that any one part of its course is better than another, but it is certain the fish get larger and more plentiful the nearer it approaches to the sea, and are caught very commonly in the salt water at the mouth of the river itself. The reason of this has been already explained. In the spring millions of small anadromous fishes begin to ascend the snow-fed rivers: on these the trout feed, and have found out that the nearer they go to the mouth the sooner this supply of food is to be obtained. Generally at the mouths of all these great rivers there is a large lagoon of brackish water. From the perfectly fresh water of the rivers they must have gone by degrees further and further into these lagoons, till at last, migrating still farther, they leave the river altogether, and are now found on many parts of the coast in the open sea, miles from any fresh water at all.

On account of the sources of the Rakaia being far back in the ranges, it is a river much subject to heavy floods, the worst being during November and December, when the snows are melting. In October, the first month of the season, the very best sport can be obtained, although the fish are not in as good condition as they are later on; but one day of northwest wind—a wind almost constantly blowing in the spring months—will bring down a flood which is almost continuous. By about January the ranges are clear, and then it is fit for the angler till the end of the season, very many heavy bags being made right up to the end of March. The angler should always make inquiry before an expedition to any snow-river.

From Christchurch or Ashburton the nearest point to the Rakaia is the station of the same name, on the main south railway line, reached in two hours from Christchurch, and an hour from Ashburton. A 14ft. rod, and a spinner of any kind, is all that is required for any part of the Rakaia, and the angler starting from the bridge can fish where he pleases. The distance from here to the mouth is about ten miles. To get to the mouth of the river, take the train from Christchurch to Southbridge—a branch line—and then take a trap of some kind to the beach, a distance of six miles. A tent is necessary if it is intended to stay more than one day, but the campiugground by the little creek running into the lagoon is perfection, being well sheltered from any violent wind, and with a beautiful spring of water, and wood in abundance. As the mouth of the river is always shifting, it is impossible to say where it will be in any year, but at no time will it be distant more than a mile from the camp. Between the lagoon and the sea a huge shingle-bank has been thrown up, and somewhere in this there will be an opening through which the river discharges its waters. The angler should find out this point and gradually fish down to it. The current of the river will probably strike the bank at some distance from the mouth, and from this point to the sea itself no better water can be found.

In these lagoons attention should always be paid to the tide, as the fish seem to come on the feed soon after low water, when the tide begins to cause the current of the river to slacken. The best possible time is generally found to be when the tide makes shortly after daylight. But even on bad days, with low tide at noon, the fish will often feed for an hour or two soon after the tide turns. At the same time the angler must not reckon too much on this. The capriciousness of our New Zealand trout is proverbial, and it is quite impossible to say that they will act the same way next year as they did the last.

A good strong rod, 13ft. or 14ft. at least, 60 yards of line, and almost any spinning-bait—the more diverse the selection the better—are essential for the mouths of all the great snowfed streams.

Some idea of the almost incredible sport to be obtained in this river, near its mouth, when the fish are really on the feed, may be gained by reference to baskets of fish made in the season of 1891-92. Before this many very heavy fish had been captured, Mr. Brice, who in 1890 was living on an island at the mouth of the Rakaia, having frequently caught very large trout, up to 18lb. in weight. On the 29th January, 1891, Mr. L. Mathias and Mr. Brice caught eight fish, weighing respectively 14lb., 14lb., 9lb., 8lb., 12lb., 13lb., 11lb., and 7lb. : total, 88lb. On the 13th February Mr. Brice, fishing by himself, took thirteen fish, weighing 147lb., the largest being 16½lb., smallest 8lb. ; 16th February, six fish, 71lb., largest 15lb. These takes, enormous as they are, were eclipsed, however, by a party of three gentlemen from Ashburton—Messrs.

Simpson, Pilbrow, and Shury. On the south side of the river the current courses along the shingle-bank, near the sea, in a similar manner to that on the north bank above described, and it was along this bank that the fish were taken. On the 30th and 31st October these three anglers took forty-four fish, weighing 347½lb., out of the Rakaia. Mr. Simpson's take weighed 173lb. 4oz., and comprised twenty-two fish, three of which were of 12lb. each, one of 10¼lb., three of 9lb., five of 8lb., two of 7lb., three of 6lb., and the remainder of 4lb. and 3lb. with odd ounces. Mr. Pilbrow's basket weighed 124lb. 4oz., and contained fish weighing 12lb., 11lb., 10lb., 9lb., 8lb., 6lb., but none lighter than 4lb. Mr. Shury, who only fished for half a day, took 50lb. weight of trout—seven fish in all, one of them 12lb.

The Rakaia has been described at some length, as what applies to it, both as regards the tackle and bait to be used and the kind of sport to be had, applies to the Ashburton, Rangitata, and the Waitaki. All these rivers are exactly similar, being affected by snow, and, with the exception of the Ashburton, equally prolific of sport and large fish. The Ashburton is the smallest of the four, and affords good sport both with fly and minnow, but the fish do not run so large as in the other three, averaging from 2½lb. to 3lb. As it runs through the Township of Ashburton, this is the most convenient starting-point for fishing, either up or down the river, or for making excursions to parts of it at some distance from the township. On this river Mr. Pilbrow killed, on the 27th November, 1891, seven fish, weighing 53½lb.; and on the 5th December, 1891, eighteen fish—110lb.

The Rangitata River is the boundary of the Ashburton and Geraldine societies' districts, and a description of it is given under the latter.

GERALDINE COUNTY.

The introduction of trout into South Canterbury streams has been attended with most marked success, and those instrumental in stocking these rivers, and thereby affording anglers such rare sport, are deserving of some little notice. As early as 1871 Mr. Melville Gray liberated trout in streams in which he was then interested. Mr. Tancred, in the same year, obtained fifty, which were placed in the Waihi, and Messrs. Gray, Tancred, and Meyer (late of Bluecliffs) released more in 1872, those of Mr. Tancred's being liberated in the upper waters of the Orari. In 1873 Mr. Gray obtained a further supply; and Messrs. E. T. Walker, R. Inwood, and J. A. Young, in the following year, commenced systematically to stock the Waihi River, now probably the best fly-fishing stream in South Canterbury. In 1875 some fine trout (*Salmo fario*), descendants of the celebrated Thames trout, were presented to the South Canterbury Acclimatisation Society by Mr. Young, of Palmerston, Otago, who had at the time a private fish-hatchery there. These were placed in the Opihi River. In 1875 Mr. C. G. Tripp released more in the Orari ; and at the same time the Hon. J. B. Acland started to stock the Rangitata. Anglers who have made themselves notorious by their takes at the mouth of this river will doubtless bear that gentleman's name in mind. In the following year some 2,500 were distributed in various streams by Messrs. W. S. Davidson, E. Cooper, Archibald, C. G. Tripp, J. B. Acland, and C. Meyer. In the year succeeding, Mr. Fulbert Archer released 300 ; and in 1878 Mr. S. C. Farr, of the Canterbury Acclimatisation Society, brought down over 4,000, which were distributed principally about Timaru. Messrs. Newton (Totara), W. Upton Slack (Pleasant Valley), and C. G. Tripp (Orari Gorge) also released fish that season. In the course of the next year Messrs. Lancelot Walker (Four Peaks), Arthur Perry (Timaru), A. McDonald (Geraldine), J. A. Gamack (Raukapuka), S. D. Barker (Kynnersley), and Tripp (Orari) also distributed their quota. In the meantime, while private individuals were busy, the acclimatisation societies were by no means neglectful of the objects for which they were organized. Trout were purchased from time to time at considerable expense; and in 1877 the South Canterbury society voted £100 for the purchase of salmon-fry, and subsequently 5,000 of Californian salmon-fry were procured by the society, which, together with a like number given by the Government, were released in the Opihi and Temuka Rivers by Messrs. Nicholas and Knight. In 1881, 10,000 trout were taken to the Mackenzie country and placed in Lake Alexandrina, Benmore, Gray's Hill Creek, and other streams, where they have grown and prospered, and, as yet, have been comparatively undisturbed. In 1885 the first English salmon (*Salmo salar*) were introduced, under the charge of Mr. Farr, the secretary of the Christchurch society. These, which were hatched from ova taken from the Tweed, in Scotland, were liberated in Temuka streams with considerable care, but it is questionable if they have thriven as it was hoped they would.

Angling for trout was, of course, for some years forbidden, but in time, as the fish were found to have multiplied, licenses were issued, and disciples of Izaak Walton shortly became fairly numerous. There were naturally many novices, and for their encouragement the Geraldine County Angling Society was formed in 1884, which had for its object the encouragement of legitimate fishing, the protection of trout, &c. Its members have been granted valuable privileges by riparian owners, and the annual competitions held under its auspices attract a good deal of local interest. The society owns handsome challenge cups presented by Messrs. A. E. G. Rhodes and R. H. Rhodes; and to have his name inscribed thereon for the best record for fly- or minnow-fishing is esteemed no small honour by the enthusiastic angler. About the same time as the angling society was started the Geraldine County Acclimatisation Society was organized, and this body has virtual charge of the fishing in South Canterbury, for in its district are included the Opihi, Waihi, Orari, Temuka, and Rangitata Rivers, with their tributary creeks; and also the Mackenzie lakes, the systematic fishing of which has yet to be commenced. The society is financially in sound condition, and has accordingly been enabled to protect the interest of anglers by employing Rangers from time to time as a check to poachers, and by keeping open the mouth of the Opihi—a river often blocked—as well as by attending to the stocking of the streams. The president of the society is Mr. A. M. Clark, of Arowhenua, Temuka, and Mr. Robert Pinckney acts as secretary, and is always willing to afford information upon all sorts of subjects within his province. Hitherto the society has issued licenses to ladies, schoolboys, and visitors at reduced rates, as well as the ordinary season-tickets, and complimentary permits are occasionally granted to distinguished visitors; but, as stated previously, it is now contemplated to issue one set of regulations for the whole of the South Island.

The Temuka

must be accepted as the centre of the fishing districts. It is a neat little township, situated upon the main line of railway, about ninety miles from Christchurch and a hundred and fifty miles from Dunedin. It has no commercial pretensions, but is surrounded by the finest agricultural and pastoral country in the colony. In its neighbourhood are several pretty private properties, and plenty of very pleasant drives may be had upon well-formed and well-kept roads. The township is fortunate in the possession of a domain of nearly 150 acres, very nicely planted. The train service from Christchurch allows a visitor to arrive at about 2 o'clock in the day. The express returns at 5.30 in the afternoon; so that at a cost of two days very nearly sixteen hours might be spent in fishing, for there are streams within ten minutes' walk of the station that upon likely days afford the best of sport. The express arrives from Dunedin daily at 5.25 in the afternoon, and leaves for the South at 2.10 p.m. There are five good hotels in the township. In the season visitors from all parts of the world may be met there, and some come regularly every year from the Australian Colonies, often staying a month or six weeks. About ten miles distant are the South Canterbury refrigerating-works, where fish can be frozen at any time. Plenty of anglers who have been successful in taking fish of unusual size take the opportunity of getting them frozen for presentation or exhibition to their friends in other colonies or at Home. Close to the Crown Hotel are two livery-stables, the one kept by Mr. Charles Story, and the other by Mr. Nicholas. The latter is a veteran fisherman, and a great favourite with visiting anglers, who are always able to obtain from him tips as to the best baits to be used and the likeliest spots to visit. As a guide he is invaluable. "Charley the Veteran" is also a very good maker of fishing-tackle, and rigs up most killing baits upon ideas of his own. There are one or two tackle-vendors in the township, whose stocks are mostly fairly complete, and tents and camping apparatus can always be obtained with a little trouble and cost. The township is within about half a mile of the junction of the Temuka and Opihi Rivers, and about five miles from the sea.

The Opihi

rises near Burke's Pass, and flows in a winding course for some sixty or seventy miles to the sea. Its characteristics are essentially those of most South Island rivers—a wide shingly river-bed with occasional patches of flax, tussocks, and toitois, intersected with rapid-flowing streams, varied here and there with deep pools. Fish in plenty are to be found throughout its course, and the angler cannot desire better sport than is obtainable in the reaches from Burke's Pass to Silverstream, a distance of about twelve miles. The first season Mr. McLeod, formerly of Burke's Pass Hotel, tried this water he killed forty-seven, of an average weight of 2lb. These were all taken with the fly. Since then records from that part have been much larger. It is a beautiful district, but not much frequented by anglers as yet, although tourists on their road to Mount Cook occasionally cast a fly with fair success. When about twenty miles of its course has been run the river enters the Opihi Gorge—a romantic spot, but not readily comeatable. Close to the western end of the gorge is the Township of Fairlie Creek, the terminus of the branch line from Timaru, and the starting-place of the Mount Cook coaches. Here there are two hotels, each furnishing comfortable accommodation. Visitors to this part will find that fish take most readily from December to February. Earlier in the season the weather is likely to be somewhat unsettled.

There are some good fish to be taken from time to time in the Opihi Gorge, but it is a spot that is only likely to attract those who combine the love of scenery with that of sport. About two miles from the east end of the gorge the Opihi is joined by the Opuha, and at about this spot there is magnificent fishing. The best way to enjoy it is to camp out for a couple of nights. This can easily be done by driving from

Temuka or Pleasant Point, which will take about two hours, the road being good the whole way. At this spot Mr. J. L. Flint took seventeen fish, scaling 87¼lb., for a night's fishing, and upon another occasion thirteen fish, weighing 61lb. Mr. Gray, at the same place, landed seven that weighed 51¼lb. These were taken in the month of January, with the minnow.

From this spot to its junction with the Te-nga-wai the river is not very much fished, although probably it is a good deal poached. There are some capital spots in its course, and fish weighing as much as 15lb. have been occasionally cast up by heavy floods. Near the junction of the Opihi and Te-nga-wai Rivers is the little township of Pleasant Point, about twelve miles from Timaru. There are two hotels here, but it has never been an angler's resort, the only people who fish being one or two of the residents. They have a very fair time of it, and have made some decent records. Close to the township is a little creek that affords nice fly-fishing. The Te-nga-wai River is probably destined to serve at some time as a "feeder" to the Opihi—that is, as regards trout. It rises in the Mackenzie Pass, and receives the waters of the Exe Creek, the greater and less Opawa, and other small streams, all of which afford capital spawning-places. On the banks of the upper part of the river is the Albury Estate, and here, under the care of the manager, Mr. E. Richardson, a number of brook-trout, supplied by the Geraldine County Acclimitisation Society, have recently been placed, and are now fit for distributing into adjacent streams.

From Pleasant Point to Temuka, a distance of about ten miles, the river is practically all fishable, and anglers upon its banks speak highly of the sport they obtain. About a mile below the railway-bridge at Temuka, the Opihi is joined by the Temuka River, and it is at this junction and in the reaches from here to the Milford Lagoon that the best fish have been taken. Trout up to 20lb. weight have been caught here, notably one taken by Mr. W. G. Aspinall; and although such fish as these are exceptional, yet the following records, all authenticated by the Geraldine County Angling Society, will serve to show that good baskets are readily obtainable.

The first published records of the society included the following:—

	No. of Fish.	Weight. Lb.	Heaviest Fish. Lb.
N. C. Nicholas	372	691½	14
J. W. Findlay	183	497	16¼
E. Gillum	172	215	5½
J. Findlay	168	364	18
K. F. Gray	46	248	16¼

These records are interesting not only to the angler, but also to the naturalist, as serving to show the phenomenal growth of trout under favourable circumstances. These trout spawn from about the middle of June to the end of September, occasionally even into October. Their natural enemy is chiefly the common shag or cormorant. Upon these latter birds the acclimatisation society have put a price, but they are so wary that very few are killed. As an instance of their voracious nature, it may be mentioned that from the stomach of one as many as fourteen trout were taken; and a competent authority ---the late Mr. W. Arthur, C.E., of Otago—estimates that the destruction of trout by only fifty shags in one year, and upon one river, allowing each bird five fish per diem (a very moderate allowance), would amount to 91,250.

To return, however, to the Opihi. At its mouth is the Milford Lagoon, separated from the sea by a sand-bank. Under certain circumstances this bank gets piled up, and the river's outlet to the sea is obstructed. At such times fishing is quite at a discount; but when the bar is opened, either by the society's employés, or by the natural action of the tides, &c., capital sport may be obtained. At such times fish bite very freely, and some very good takes are made. Here are one or two: Mr. Fitzgerald and party (3 rods), eight fish, scaling 40lb.; Mr. Flint, nine fish, 46½lb.; Mr. Nicholas, eleven fish, 46lb., and again, five fish, 30½lb. In addition to trout, kahawai and grey mullet may also be caught; in fact, with favourable weather, the Milford Lagoon and the lower reaches of the Opihi are very enjoyable spots to visit. On Green Island, near the mouth, an old fisherman has established himself, whose assistance sometimes proves of value to the stranger, and although boats are not regularly on hire, yet the use of one for a few hours can generally be obtained if desired. Here is another record of a season's fishing almost exclusively in the Opihi. (Although the name is not mentioned, the record has been certified to.)

	Fish.	Lb.
October	84	254¼
November	96	298¾
December	86	327
January	106	385¼
February	96	382¼
March (to the 18th)	39	150¾
Total	507	1,799

In the best day's fishing the average weight was $6\frac{2}{16}$lb.

The Temuka River

sweeps in a semicircle about the township, and is reached from any part in about five minutes' walk. It affords for some three miles capital fly-fishing, but it is not so frequented now as formerly, the attraction of the larger rivers, with their heavier yields, proving too much for the ordinary angler. It has, however, a number of consistent admirers, anglers who know its every ripple and shoal, and who, in the early mornings and evenings, enjoy many a pleasant hour, and catch many a lusty trout. It is a good spawning stream, and generally contains plenty of feed. Trout are certainly not so plentiful in its lower reaches as formerly, and this may be attributed to one or two causes. The river's propinquity to the township is at once an advantage and a disadvantage. It is practically all open water (although riparian rights are claimed by one owner), and, as a consequence, it has in the past been somewhat over-fished. Access to its upper waters (the Waihi) is partially obstructed by a mill-dam, the property of Messrs. Aspinall and Co., and, as a consequence, certain of the best spawning-beds are inaccessible, and the supply from the upper part is, in its turn, shut off. The question of having a close season over a few miles of this stream has been debated, but it is probable that the acclimatisation society will prefer to re-stock.

Just above and below the traffic-bridge which crosses the stream are two very nice ripples, where a fish can almost always be obtained, and a little higher up, between Myers's willows and the Manse foot-bridge, there is also a good spot. All the way from the foot-bridge to the mill the fishing is fair, there being one or two exceptionally favourite spots. That it is still a favourite place with many people may be partially attributed to its scenery. Anglers as a rule are admirers of Nature, and to those who stroll along this river-bed not too intent upon sport the view possesses many beauties. The barrenness of the ordinary river-bed gradually disappears as one ascends the stream. Here and there the banks are more or less clearly defined, vegetation is comparatively luxuriant, and on either side glimpses of well-kept farms with orchards and plantations may be obtained. In places, willow protective-works have been erected, and the foliage of these in the early summer affords a pleasing change to the ordinary sombre tints of the pines and blue-gums. As a background there are the splendid ranges of Four Peaks and Mount Peel, which in the singularly clear atmosphere of New Zealand appear only a few miles distant.

Just below the mill the river receives the Hae-hae-te-moana, which in turn a few miles higher up has absorbed the waters of the Kakahu. The former is scarcely ever likely to be a reliable stream for the angler. Its course is very short, and in dry seasons its waters are almost lost in the shingle. The Kakahu on the other hand is never dry, and there are some really capital places in its upper course as yet practically unfished. In the course of time, unless poached, this stream will be a favourite for fly-fishing. Above the mill the river for a few miles is really private property. The owners, Messrs.

Aspinall and Co., McCallum, Twigg, and Cliff, reserve to themselves the right of fishing; but none of them are very chary in granting an occasional day—all have accorded to the members of the angling society the right of fishing these streams on competition days. As an instance of the sport obtainable in the Temuka River with the fly only, the following records kept by Mr. W. Mendelson are given: Season 1889-90, for the months of November, December, January, February (6 days), and March (1 day): total caught, 536; of which 100 were put back as being under size. The remainder weighed 514lb. 15oz. The following season, his record from November to February was 413, weighing 341¼lb. In addition, 149 were put back. Individual baskets were: eighteen fish, 17$\frac{13}{16}$lb.; fifty-five fish, 22lb.; five fish, 6$\frac{2}{16}$lb.; eleven fish, 11$\frac{9}{16}$lb.; twenty-four fish, 11lb.

Mr. Mendelson was the winner of the angling society's prize, and Mr. R. H. Rhodes's challenge cup, in 1890; his take being fifty-five fish, 27lb. Other records upon that occasion were: Mr. J. Findlay, nineteen, weighing 12½lb., and Mr. Nicholas, ten, scaling 15lb. These, it must be remembered, are the takes for probably less than eight hours' fishing. At the last competition, the 16th December, 1891, the challenge cup and first prize was awarded to Mr. N. C. Nicholas for a basket containing forty-five fish, weighing 33½lb. Mr. W. Mendelson made the best average, his take being seventeen fish, 22⅞lb. Upon this occasion no fish under 10in. were allowed to be weighed. Favourite flies are the "Guv'nor" (small), "red spider," "Hardy's favourite," "blue uprights," and "March brown;" but, of course, a great deal of the angler's success depends upon his skill and observation.

Above Mr. Twigg's farm the immediate confines of Winchester are reached. Here we come to

The Waihi,

which is a fair fly-fishing stream. It is a favourite resort of anglers, and indeed of visitors generally. The river is very pretty, with its nicely-grassed banks, drooping willows at intervals, and generally peaceful and beautiful surroundings. It rises in the neighbourhood of Woodbury, some sixteen miles nearer the hills, but its upper waters are not fishable. For about three miles above and below Winchester the best sport can be obtained, though the water near the hotel is much fished, and anglers will do best by getting a mile or so away. A certain portion of the water is preserved, but, by arrangement, the lessee of the Wolseley Hotel, Mr. Langdon, has acquired for his customers liberty to fish all the best of the water. Mr. W. De Renzy, of Hunnington, has riparian rights over some four miles of stream, but he does not exercise them to any extent; that is, if asked for permission to fish he accords it, unless it be that he has reason to think too much pot-hunting and poaching is going on.

The little village of Winchester, in early days known as Waihi Crossing, is a pleasant little spot with a peculiar appearance of the Old Country about it. It contains two hotels; but "the Wolseley" before mentioned is *par excellence* the "Anglers Rest." A glance through the visitors' book shows what a number of genuine anglers have stayed there, and all make favourable mention both of the accommodation and sport. Here are a few records from this part, taken haphazard, but authenticated: Fourteen fish, $18\frac{1}{2}$lb.; twenty-eight, weighing 32lb.; eighteen, averaging 1lb.; thirty-nine not weighed; eighty fish for two rods in three days; twenty-two fish, 14lb.; twenty-five fish, 16lb.; twenty fish, 15lb. Winchester is easily accessible by railway, it being eighty-four miles from Christchurch on the main line. Latterly the express has not stopped there; but visitors can always get off at Orari, and, by telegraphing to the proprietor of the Wolseley Hotel beforehand, they will be met with a dog-cart or other conveyance and brought on.

The Orari River

runs almost parallel to the Waihi, about four miles further north. It is a river that is not often fished, the best parts of it being either somewhat inaccessible or else preserved. The upper part, near the Orari Gorge (Mr. Tripp's property), is fairly well stocked, and anglers who have had the opportunity of fishing it speak well of their takes. Its position is such, however, that it scarcely repays a visitor unless the visitor desires a change of scenery as much as sport. Nearer the mouth, at which there is a small lagoon, sport is better, and this place is fished pretty regularly from December to February. A tributary of the Orari is the Ohapi Creek, a splendid little trout-stream, from which many a basket is taken with the fly. Mr. Guild of Trevenna, Mr. Gladstone, and other owners are not shy of granting permission to anglers who desire to fish the portions flowing through their estates, but, believing that what is worth having is worth asking for, object to unauthorised visits.

Of late years the heaviest baskets have been taken from

The Rangitata,

the most northern stream in the Geraldine County Acclimatisation Society's district, and which they share with the Ashburton society. It is a splendid stream, teeming with fine fish, and offering irresistible attractions to those anglers who can afford a reasonable time to visit it. It is distant from Temuka or Winchester (the best places from which to start) about fifteen miles—that is, to the mouth of the river, and the general method adopted by visitors is to secure the services of Mr. C. Nicholas to drive them out, and start them upon the best water. Mr. Nicholas has a convenient accommodation hut, with all necessaries, near the mouth of the river, and also a couple of handy punts. Here a day or two can be pleasantly spent, and the sport is almost certain to repay a visit. Occasionally, of course, a trip may be made without result, for the river, under the influence of a nor'-wester, rises very suddenly; but, as a rule, information as to its condition is obtainable, and unless this was favourable the angler would not of course start. Another plan of visiting this place would be to hire a conveyance, with tent and blankets, and select one's own camping-ground. Many regularly do this from choice, and more from the fact that it requires early application to get the use of the hut. Once upon the ground, there is every prospect of sport, for there is generally some one about in the season who knows the run of the river, apart from the fact that an angler with reasonable experience can hardly go wrong in his selection of his casting-ground. Very few people fish far up the stream, although there are plenty of good spots. The reason of this is that, as with most other snow-rivers, the nearer you get to the mouth the better. Decent fishing may be obtained, however, by anglers starting from the Star of the South or the Rangitata Bridge Hotels. Both of these are near the stream, and can be reached with comparatively little difficulty. Probably the reason anglers do not frequent them more is the dread of the loneliness of a by-day when the river is unfishable, and there are no neighbours within a few miles.

On the opening day of this (1891–1892) season very good sport was had at the mouth of the river, although the fish were in rather poor condition. Mr. Nicholas caught twelve that weighed 53lb.; Mr. Meredith Kaye four, weighing 30lb.; Mr. Hope seventeen, scaling 81½lb.; Mr. Hollwell four, weighing 21lb. On the Ashburton side of the river Mr. Huston took nine, scaling 53lb.; Mr. W. F. Somerville ten, weighing 50lb.; and there were several smaller baskets.

Here are other records from the same river during the same season:—

	Rods.	Fish.	Lb.
September 26 (2 hours' fishing)	1	6	42
" (4 hours)	1	13	78
October 9 (1¾ hours)	1	11	87¼
" 15, 16, 17 (7 hours)	3	30	269½
" 26 (4 hours)	1	9	78½
December 8 (6 hours)	3	16	178¼
January 19	1	5	52
" 28	1	9	54½
February 10	1	5	37
" 19	1	9	51
" 26	1	7	52
March 7	1	14	96½

The largest fish of the above weighed 19lb., and all were taken with the Devon minnow and whitebait phantom. The natural bait was not used. The above records should serve to show that the fishing at Rangitata mouth is some of the best in the colony.

Another spot, destined at no distant date to come into popular favour, is Tekapo Ferry, in the Mackenzie Country. From this centre some of the best creeks and rivers can be fished as well as the lakes. Lake-fishing, as far as the Mackenzie Country is concerned, is still in its infancy, no one as yet having attempted trolling from a boat. It is probable that after a time the lakes will be netted, but this cannot be successfully undertaken until suitable provision is made for the temporary storage of the trout in ice or otherwise, and their transmission to a suitable market. This is certainly to be done in the no-distant future, and in the meantime the fish are growing and multiplying. As a general note in connection with the whole of the district included in the Geraldine County it may be mentioned that the roads are remarkably good, conveyances fairly comfortable and well horsed, and the hotels well kept, and the charges by no means excessive.

The South Canterbury Society's district is bounded by the Opihi, already described, and the Pareora. The Pareora runs into the sea about nine miles south of Timaru. There is occasionally fishing to be had in this river some ten or twelve miles up from the sea, and about fourteen or fifteen miles from Timaru, when, after a wet season, a few biggish fish come up the river to spawn; but the sport is so uncertain and usually so indifferent when obtained that it is not worth while entering into details.

WAITAKI AND WAIMATE DISTRICT.

This district comprises a large area of country in the South Island. It extends from the Pareora River on the north to the Kakanui Mountains on the south, a distance of some fifty or sixty miles; its western boundary, roughly speaking, being the Alpine Range in the vicinity of Mount Cook. The main stream in this district is the Waitaki, one of the largest of the great snow-rivers.

What applies to the Rakaia applies also to the Waitaki. Throughout its length it is abundantly stocked with trout of enormous size and excellent flavour.

Three large streams—the Ohau, Pukaki, and Tekapo—coming from the highest mountains in New Zealand, and each passing through large lakes, combine to form the Waitaki River. Two other smaller streams—the Ahuriri and Hakateramea—join it lower down, the first named being the beau ideal of an angler's stream, either for fly or minnow, and both, being out of the reach of the ordinary traveller, are very heavily stocked.

The Oamaru Creek, a tributary of the Ahuriri, is also well stocked. The Otematata Creek is stained with water from the Mount Burster Gold-diggings, and it is of no use to the angler, except at such times as when from the want of water the diggings are not used; then the trout go up from the Waitaki.

The Hakateramea runs into the Waitaki below Kurow. It is a beautiful clear stream, and affords sport with the fly, and in the early season minnow, many large baskets having been taken out of it.

The Maerewhenua, which also runs into the Waitaki, is of no use for fishing, being always polluted by the washings from the Livingstone Gold-diggings; the result being that it is always like pea-soup, and any trout caught near its mouth are thin flabby fish, quite starved.

For the whole of the Upper Waitaki and its tributaries, Oamaru should be made the starting-point. This can be reached either from north or south, it being one of the chief stations on the main south line. A branch railway runs from Oamaru to Kurow and Hakateramea, a distance of forty-three miles. If it is desired to try the Ahuriri, stop at Kurow and obtain a vehicle of some description. A tent and camp-equipage also should be taken.

The Upper Waitaki can be fished either from Kurow or lower down at Duntroon. Here there are two hotels, but leave can be obtained to camp, if desired, in an orchard, within five or six minutes' walk of the river. The minnow is the best bait for the larger fish, the favourite being the Devon (size, 1½in.). On a calm still evening they will take the fly freely. Lake-trout size fly is the best, the fish running up to 4lb. or 5lb.

The main line between Christchurch and Dunedin crosses the Waitaki about thirteen miles north of Oamaru. On each side of the bridge, which is a mile long, is a station—Waitaki North and South—with an hotel at each. By staying at either of these (perhaps Waitaki North for preference), the angler is close to the river, and only distant about two miles from the mouth, where sport is to be obtained not to be surpassed. Many very large and heavy baskets have been taken out of the Waitaki between the bridge and the sea, 80lb. and 90lb. weight to a single rod being not at all uncommon for a single day's sport. This part of the river swarms with large and very powerful trout, yet I made here once the worst record I know of. The largest fish that has been caught was taken by Mr. Henry Aitken, and weighed 21¼lb. This is one of those huge rivers where trolling with the minnow (and for the lower part nothing else is of any value), close to the side, is nearly always successful.

The Kakanui River is about eight miles south of Oamaru, and is a clear stream full of fish, and affords the best fly-fishing in the district, large baskets have been taken out of it both with fly and minnow. Dr. de Lautour has taken fish of 7lb. weight with the fly, and on one evening took five fish weighing 36lb. with the minnow, the heaviest being 16lb. Mr. Alexander Thomson also, in the season of 1891-92, one evening caught eleven trout weighing 50lb.; his companion (Mr. H. Searle) caught at the same time seventeen trout averaging $6\frac{1}{4}$lb. The Waianakarua stream, about twenty miles south of Oamaru, also affords good fly- and minnow-fishing; it was much affected by this last dry season, and quantities of fish perished, but it has since been restocked.

The Waihoa River in South Canterbury, about seven miles north of the Waitaki River, is easily accessible from Waimate or Waitaki North Railway-stations, it is well stocked with trout, and anglers residing at Waimate speak highly of the sport to be obtained in that river.

In addition to these streams, various other slow-running streams, lagoons, and ponds have been stocked with perch and tench, and have thriven well—perch weighing 4lb., and tench over 3lb., have been taken.

OTAGO.
The Clutha and its Tributaries.

The Middle Island of New Zealand is only about 550 miles long, with an average breadth of 120 to 124 miles, and contains a little over 55,000 square miles; it is, therefore, surprising to find that it contains a river of such immense volume as the Clutha River, or Molyneux, as it is sometimes called. It is stated that this river discharges into the sea an average all the year round of 1,690,400 cubic feet of water per minute. Of the Thames the discharge is only about 102,000ft., whilst that of the Clyde is no more than 48,000ft.; so that this Otago river pours forth a volume of water about sixteen times greater than the Thames, or nearly forty times greater than that of the Clyde, and discharges about 300,000 cubic feet per minute more than the Nile.

The Clutha takes its rise in the Wanaka and Hawea Lakes, and it also drains Lake Wakatipu, through the Kawarau River, which joins it at Cromwell. These three lakes have a combined area of about 240 square miles, and it is mainly on account of this wide expanse over which the flood-waters, caused by the melting of the snow, have first to spread that disastrous floods in the lower basin of the Clutha do not more often occur. The total length of the Clutha proper, from where it leaves Lake Wanaka, is 150 miles, and its principal tributary streams, besides the Kawarau, are the Manuherikia and Pomahaka. Like all rivers which derive most of their water-supply from the melting of the snow in the high lands, the Clutha is very low in winter during cold and frosty weather, and keeps at a high level all summer—warm weather in early spring, which melts the snow suddenly, being greatly dreaded by the settlers on account of the floods it sometimes causes in the lower basin. The general direction of the river is from north to south, though in some places it winds a good deal, the upper portions being for the most part rock-bound and the channel narrow. In its lower reaches the Clutha is from two to three hundred yards wide, and runs at the rate of from two to five miles per hour; but higher up, where the channel is more confined, the rate of the current in some of the gorges is as much as ten miles an hour. In spite of this and the danger from sharp rocks and snags, the river has been twice navigated by boat for a considerable distance—about thirty years ago by Messrs. Hartley and Riley and a crew, as far as the Dunstan; and last year by Mr. H. J. Day and Mr. Pillans, from Balclutha to Roxburgh and back. It took the latter nine days hard work to go up, and they returned with ease in ten hours. The Clutha is navigable by small steamers for about forty miles from its mouth. Between the mouth of the Kawarau and the mouth of the Pomahaka its waters are much polluted by mud and tailings from the gold-diggings, and are not worth fishing; but, with the exception of this stretch of about ninety miles, the Clutha and its tributaries afford fishing-waters to the angler which, for their extent, and the quality, size, and abundance of trout they contain, are unsurpassed even in New Zealand.

The lower basin of the Clutha and its tributaries from the Pomahaka to the sea are much the most important to anglers from Dunedin or other distant places, in consequence of their accessibility from the numerous stations on the main south railway-line. I therefore propose to begin at the mouth of the Clutha and gradually work up stream, giving the reader some notes of the fish and fishing, and such information regarding localities as may be interesting or useful to the tourist-angler or visitors from a distance.

The Clutha discharges its waters into Molyneux Bay by two mouths—the old one near Port Molyneux, and the new one, which was burst by the great flood of 1878, through the sand-spit near Coal Point, at the north end of Molyneux Bay. About nine or ten miles from the mouth in a direct line is situated the Town of Balclutha, near which the river divides into two branches; the westerly one, called the Koau, running almost direct to the sea, the Matau or easterly branch running in a tortuous course past the coal-mining Township of Kaitangata. These branches join before reaching the sea, forming by their division the fertile island of Inch-Clutha, around which there is some of the best fishing-water in the river. As often mentioned before, there is no reason to doubt that many of the trout go out to sea, some of them probably going in and out of the river with the tide, which rises and falls about 4ft. to 5ft. at the lower end of Inch-Clutha, and, though it does not much impede the flow of the river, has a slight effect on its level even so far up as Balclutha Railway-bridge. The trout in these waters live almost entirely on the small migratory fish which from October to April swarm into the mouths of the river in immense shoals. These shoals run up the river close under the banks, and the trout lie in wait for them behind each projecting point where the water makes a ripple. It is, therefore, useless to cast out into the river, this habit of the fish having caused the plan of trolling close to the bank with a short line to be largely practised, even as far up as the mouth of the Pomahaka. The trout weigh from 2lb. to 7lb., and have been taken up to 14lb. and even 18lb. in weight. On more than one occasion a bag has been made of sixteen trout by one rod in one day, weighing over 60lb.

The first tributary we meet with in our passage up stream—the Puerua—enters the Clutha near Port Molyneux, and has a course of about fifteen miles fishable water. The lower reaches are tidal, and though they contain many and large trout, are difficult to fish except from a boat; but in the upper waters some excellent bags of large fish are taken, particularly about the middle of the season, with cricket. There is a comfortable accommodation-house near Puerua Bridge, distant nine miles from Balclutha, on the Port Molyneux road. Nine years ago a trout weighing 22lb. was caught near this hotel, this being the largest of which any record has been kept for the Clutha. There is also a creek which runs into the Puerua, called the Glenomaru Stream. In the spring the trout in this creek take the artificial fly very freely, and some large baskets were taken last season. The fish in the Glenomaru, however, do not run very large—2lb. and under, averaging about 1lb. Near the railway-station at Pomahaka is an accommodation-house, where the angler can stay while fishing this stream.

The next tributary of the Clutha flows into the Matau branch at Kaitangata, and is called the Kaitangata Creek. It drains the lake of the same name, and also, through Morrison's Creek, the Tuakitoto Lake. These are large shallow lakes, nowhere more than 6ft. deep, and averaging about 2ft. They contain large trout, and some excellent fish have been obtained at night by spinning with a whitebait phantom in the above-named creeks, as also in the lower waters of Lovell's Creek, the principal feeder of the Tuakitoto Lake; but for the last three years these creeks and lakes have literally swarmed with perch (*Perca fluviatilis*), to a considerable extent spoiling the trout-fishing, though the perch-fishing is most excellent, as many as 175, weighing from ½lb. to 1½lb., having been taken in six hours by two rods, spinning with artificial minnow.

In the upper waters of Lovell's Creek the trout take the fly well in the forepart of the season, Lovell's Flat Railway-station being distant about nine miles by rail from either Milton or Balclutha, where excellent hotel-accommodation can be obtained by the angler.

To fish the lower parts of the Matau, the angler should stay at Kaitangata, where there are two good hotels. The best place to put up at in order to fish Morrison's Creek, Tuakitoto Lake, and the waters which surround Inch-Clutha is the Railway Hotel at Stirling, both of these townships being easily accessible by rail from Dunedin.

The next tributary of any size is the Kaihiku, which runs into the Clutha on the west side, about six miles above Balclutha, the upper waters of which can easily be reached by rail, the angler getting out at Kaihiku station, on the main south line, about twelve miles above Balclutha. This stream has about one mile and a half of sluggish backwater at its mouth, where trout and perch may be captured by trolling behind a boat, or from the banks on a windy day, and the upper waters contain good gravel-reaches, which form a splendid spawning-ground and nursery for trout, some very good baskets having been taken there every season with the artificial fly, particularly during the latter half of October and the month of November. The average weight, however, is not much over 1lb., because, the stream being small, the larger fish drop down into the sluggish water at the mouth and eventually into the Clutha River. Indeed, this stream may be considered one of the best feeders of the Clutha so far as the supply of trout is concerned. To fish the Kaihiku and the Koau branch of the Clutha, and also that river as far as the mouth of the Waiwera, the angler cannot do better than make his head-quarters at Balclutha, where there are four good hotels, and every facility for hiring saddle-horses or buggies. It is also within easy distance by train to Lovell's Flat, Stirling, Kaitangata, Kourahapa, and Kaihiku. Indeed, this township, besides being the centre of a large and important agricultural district, is also the centre from which the best fishing-waters of the lower Clutha basin are most easily accessible.

The next tributary of the river with which we meet—the Waitahuna—comes in on the east side, and for about eighteen or twenty miles of its course—that is, from the Waitahuna Township, or Havelock, as it sometimes called, to the mouth—this stream is so polluted by the diggings, that its waters resemble pea-soup. It is needless to say that no angling can be obtained in this stretch of water; but some little distance above the township, where the stream is free from tailings, many fair baskets have been taken with grasshopper, cricket, and artificial fly, though the average weight of the fish is not very high.

The next two streams we have to consider—viz., the Waiwera and the Pomahaka — join the Clutha within a few hundred yards of each other on the west side. The mouths of these rivers have the same characteristics as that of the Kaihiku, having, the former one and the latter, about two miles of sluggish backwater at their mouths, which abound in trout, perch, and also huge eels.

On the lower waters of the Waiwera leave should be asked for from the proprietors who own the land on both sides of the stream, but above where the main south road crosses the Waiwera there is excellent fishing-water, where there are annually recorded large takes of good fish. The Waiwera trout take fly well, the cricket and black-creeper being also very deadly baits.

The Kuriwao, an affluent of the Waiwera, which joins it after passing near Clinton Township, helps much to keep up the reputation of its sister-stream. It has excellent gravel fords, and is one of the grand stand-byes of the rangers of the Otago Acclimatisation Society when the season arrives for collecting ova for their trout-breeding establishment on Marshall's Creek, near Clinton.

To fish the Waiwera, Kuriwao, and also the lower waters of the Pomahaka, the angler ought to make his head-quarters at Clinton, distant seventy-four miles from Dunedin. There are two good hotels, and means of conveyance to the streams in the neighbourhood can be easily obtained.

Before proceeding to treat of the Pomahaka and its tributary creeks, a few more words may be written about the fishing in the Clutha itself. The banks below Balclutha are chiefly composed of alluvial clay, but above the township the angler will find rocky points round which to spin his minnow. Good trout, running from 2lb. to 6lb., and even as high as 10lb. in weight, are to be caught from Balclutha as far up as a mile above the mouth of the Pomahaka River. One of the best bags known of as having been taken in this stretch of water by one rod in a day was thirteen trout, weighing 53lb. 5oz., the largest being 8lb. 12oz., and the smallest 2lb. 12oz.

The Clutha trout of both sexes acquire a white colour and lose the red spots usually to be seen on those which inhabit the creeks, retaining only the black spots, and it is thus easy to detect by its colour a trout which has recently dropped down out of one of the tributary streams to take up its abode in the river. The trout caught near the mouth of the river are of a very bright silvery appearance, with few spots, and those black, and generally of a cruciform shape. Occasionally one is caught of a golden-yellow colour. The creek trout are usually of a darker hue, and the males have red spots, more or less bright.

Sometimes trout taken in the beginning of October are not in very good condition, but after the middle of November very few ill-conditioned fish are caught; and towards the end of the season they become very fat, and so shy that it is difficult to take them, even with the most tempting natural bait. The Clutha River trout, particularly those caught near the sea, are excellent eating, being entirely free from the muddy or mossy taste which so often spoils the edible qualities of those caught in the creeks. The flesh is of two distinct colours—some have pink or salmon-coloured flesh, and some yellow. I have been unable to detect that this could be accounted for by the sex of the fish, though it is just possible that it may have something to do with the age. There is great difference of opinion as to whether the pink- or yellow-fleshed trout are the best eating, but generally the latter are firmer, more flaky, and better flavoured.

There is a chain-wide reserve along the Clutha River for the greater part of its course, and in many cases half a chain along the creeks in the district, so that the angler is free to fish without trespassing; but in some places the Government reserve along the river-bank has been washed away, and in other cases the creeks have been surveyed in with the sections and sold, thus becoming the absolute property of the owner of the land. This has, however, never yet led to much trouble or any action being taken by the owners of the land against anglers. The river-side settlers are kind and hospitable, and it is most unusual to meet with one who would refuse to allow the angler to fish along his frontage—provided he asks leave civilly, and does not leave gates open, break gaps in the fences, or bring dogs which disturb the sheep. Many of them are fishermen and true sportsmen who would gladly give a visitor advice about the best places, baits, and times of the tide, and other local hints, which help so much towards the success and enjoyment of the angler.

Now, proceeding on our way up that splendid trout-river, the Pomahaka, of which there is over sixty miles of good fishing-water, we find between the lower and upper Clydevale Bridges grand rugged rapids and rocky pools. Above the upper bridge the stream winds through a strip of low country, called the Burning Plain, on account of a seam of lignite being on fire, which has evidently been smouldering for a great many years. The Pomahaka here runs over reefs of lignite, which have been hollowed out in many places by the current into deep gutters, in which trout lie in wait for minnows. Trolling with a long line is here practised, the fisherman throwing his bait far across the stream, letting it come in towards him in a semicircle. Several prize baskets have been taken out of this stretch of water with artificial minnow, but a few good trout are also taken with fly in the spring and cricket in December and January. For minnow-fishing the lower reaches of the Pomahaka may be considered a late water, partly on account of its being to some extent snow-fed; the trout being in the best condition and taking best after the middle of January. The upper Clydevale Bridge is distant about ten miles by road from Clinton. Following up the Pomahaka we find a series of rocky gorges, affording to the angler alternately rapid water and long still pools on which to try his skill. These are well stocked with trout from 2lb. to 6lb. or 7lb. in weight; the Back Creek, Rankle Burn, and Wairuna, being all good breeding-streams, help to keep up the stock in this stretch of the Pomahaka.

We now come to the most important tributary of the Pomahaka—the favourite stream of Dunedin fly-fishers—the Waipahi. This is one of the very best trouting rivers in Otago, and is a stream that the trout love to tumble and leap in. Its waters are as sweet and wholesome as the nectar the gods partook of. There are deep rocky lairs all overgrown with long slender weeds that stretch their feathery fronds in all directions, affording splendid hiding-places—nooks that a mermaid herself would covet. How calmly and placidly it meanders down the valley when it leaves the rugged rocks and tussock-covered hills! It never seems in a hurry to join the rocky snow-fed waters of the Pomahaka. When the Otago angler needs rest in invigorating exercise he hies him away to this stream, and drinks in the pure ozone of the hills, meanders along its banks and lures the finny denizens from its crystal pools, thus viewing nature in all her varied forms. Although the trout in this stream are inclined to be shy and turn up their noses at the lures placed so temptingly before them, yet when they are on the take they give more sport than almost any other; and for delicacy of flavour they are not to be surpassed, the flesh being of a delicate pink colour with layers of creamy curd between each flake of flesh. This, though not a very good breeding-stream, may be considered one of the earliest fishing-waters in this part of the country, as it is not snow-fed, and is low-lying and well exposed to the sun. Almost weekly throughout the season reports may be seen in the angling column of the *Otago Witness* of the splendid baskets of grand trout taken from this stream. It is also generally chosen as the river on which the annual fly-fishing competitions are held. Near the Waipahi Railway Station (distant eighty-four miles from Dunedin) there is a good hotel, the favourite resort of the anglers. From here the visitor can make fishing excursions by rail, either to Arthurton, on the Waipahi, or to one of the small railway-stations on the Tapanui line, in the valley of the Pomahaka River, above its confluence with the Waipahi. The Waipahi has one important tributary, called the Kaiwera or Otaria Creek, which may be fished from the Otaria Hotel, distant fifteen miles by road from Clinton. The upper waters of the Waipahi also afford good sport, and may be reached by the road from Clinton to Otaria, distant nine miles from Clinton.

Proceeding up the valley of the Pomahaka, we come to what may be called its head-waters, and the numerous creeks which run into them which may all be fished from—Tapanui, Kelso, or Heriot—at each of which places the angler can get good accommodation within easy distance of good fishing-water. The principal streams are the Crookston, Heriot, and Spylaw Burns, and the Leithen Stream, besides a number of small creeks which, together with the Pomahaka, form a pleasant fishing district, where an angler may spend a few weeks with every prospect of good sport.

There are several good-sized streams running into the Clutha, between the mouth of the Pomahaka and Cromwell, which, should the angler find himself in their neighbourhood, are well worth a trial, as fair baskets taken from them have been recorded; but it would not be worth his while to make a special journey to fish there, more particularly as they are out of the line of railway communication. They are the Blackcleugh and Earnscleugh on the west, and the Beaumont (near Beaumont Hotel), Talla, Fruid, and Minzion Burns on the east side; and further up on the same side we have the Teviot River, and the Manuherikia, and its numerous tributaries. In all these streams trout have been liberated, and, where the waters are not polluted by tailings, or diverted for sluicing purposes, they have thriven well and increased.

In the next stream we come to, the Kawarau, which drains Lake Wakatipu, there is no fishing, as its waters are very thick with tailings, particularly below its confluence with the Shotover—another diggings river, little better than a sludge channel. The principal streams which run into Lake Wakatipu are the Locky, Greenstone, Dart, and Rees Rivers. These streams are under the control of the Lake Acclimatization Society, by which they have been thoroughly stocked with trout. The lake trout are generally fat and short in proportion to their weight, running from 2lb. up to as high as 30lb. in weight, annually ascending the above-mentioned rivers to deposit their spawn. They are, nevertheless, so difficult to catch in the lake with rod and line that the society has of late years let the shores of the lake in sections for netting, and, as the fish keep moving back and forward along the sides, where the water is shallow, large numbers are annually captured in this manner, finding a ready market in Queenstown, Invercargill, and Dunedin. Lake Hayes, between Queenstown and Arrowtown, has for the same reason been let for netting. Anglers visiting Glenorchy, at the head of the lake, ought not to lose the opportunity of trying the Rees River, particularly near its mouth, and after dark. Trout are sometimes taken in the lake by trolling with a long line behind a boat in the evening.

We have now to consider the fishing on the Upper Clutha. From two miles above Cromwell up to Albertown, a distance of about forty miles, there is first-class fishing to be obtained on either side of the river. A coach (Cobb and Co.'s) runs three times a week from Cromwell to Albertown, and there are three hotels on the road—namely, at Lowburn (about three miles from Cromwell), and the Queensbury and Luggate Hotels (about twenty and thirty miles respectively from that town). All these hotels are near the river, especially the two latter, and most conveniently situated as resting-places for the angler while fishing the best reaches on both sides of the river. Cobb and Co.'s coaches run along the west side of the river between Cromwell and Luggate, so that on the east side the conveniences for tourists are not so good, there being only one hotel, which, however, is centrally situated for a large extent of excellent fishing-water: this is Rocky Point Hotel, about ten miles distant from Cromwell.

The principal tributaries running into this stretch of the Clutha are: the Lindis, which has been so much poached that it is scarcely worth fishing, and the Hawea, which joins the Clutha at Albertown and drains the lake of the same name. There are good trout to be got in this stream, and also in the Clutha near the last-named township, which possesses a comfortable hotel. The means by which anglers can cross the river are by punts at Lowburn, Luggate, and Albertown, and by a chair suspended on a wire-rope, near Queensbury Inn. I am informed by a local fisherman of great experience, to whom I am indebted for much valuable information regarding the Upper Clutha, that although good sport may be obtained on almost any part of the river above Cromwell, yet he especially recommends those portions near Queensbury and Luggate Hotels, fishing the west side in the afternoon till dark, and the east side in the morning from day-break till the sun gets too high — these being the only times of day when the fish take freely; a brisk breeze from the south and a dull day being the best sort of weather. The angler ought not to visit this part of the Clutha before the middle of December. By that time the snow-water is all down and the fish well on the feed. The minnow is here considered the most deadly bait; indeed, any other lure is seldom used, though a few fish are sometimes taken with grasshopper and cricket. In the neighbourhood of Luggate the average weight of the trout is about 2½lb., but the higher up the river you go the more the average weight increases—an occasional fish being landed that turns the scale at 12lb. Much heavier trout are often speared, some thus poached weighing over 20lb.

The best baskets I have heard of were taken near Luggate —this season, sixteen trout weighing 40lb., and last year eighteen trout weighing 48lb. These were taken by one rod in four hours' fishing.

Lakes Hawea and Wanaka are full of large trout, but I have been unable to hear of any having been taken with rod or line. Whether this is caused by the fish altogether refusing to take in the lakes or by the inexperience of those who have tried, I cannot say.

There are several considerable streams which run into these lakes, and they are doubtless inhabited by trout, but I cannot find any record of their having been fished.

There is no doubt that there are streams in New Zealand which afford to the fisherman trout of a greater average weight, but the tourist-angler may rest assured that for the extent of its fishing waters, the gameness and condition of its trout, combined with the beauty, variety, and in the upper basin the grandeur of its scenery, few rivers equal the Molyneux or Clutha River and its tributaries.

The Clutha and its tributaries comprise three-fourths of the best fishing waters of Otago; there are a few other rivers, however, which afford excellent sport, the first of which mention is to be made is the Taieri River.

The Taieri River takes its rise in the Naseby district, about a hundred miles from where it flows into the sea. Unfortunately, it and its tributaries are all more or less affected by the diggings, excepting its upper waters, where there is splendid fishing to be had. There being no railway communication the angler has to go by coach from Middlemarch or Palmerston. Good hotel accommodation is to be had within reasonable distances from the river in many places.

Four miles from the mouth of the river is the Waihola Lake, full of fine fish, which are taken by the minnow on a very calm day; but should there be a breeze on the water becomes too thick for fishing. Eighteen miles up the Lee Stream comes in, extending westwards for sixteen or seventeen miles. At one time this was Otago's best stream; but, alas! a rascally Chinaman found gold in its upper reaches about ten years ago, consequently the anglers of Dunedin had to seek fresh ground. It is a very nice stream to fish, the lower reaches being one series of gorges with fine pools; ten miles of the higher waters having level banks with grand runs and pools, making Dunedinites wish for the old days back again. This year it has been clear for about half the time, which makes one hope the diggings will soon stop.

Eight miles further up the Deep Stream joins the Taieri, and the same remarks apply to it as to the Lee Stream, only the stream is about a fourth larger.

Twelve miles further up the Sutton comes in; it is much smaller than the Lee or Deep Streams, but is clear, and contains good fish; is very little fished on account of being situated in an out-of-the-way place; it extends westward about ten miles. The best baits for these streams are the grasshopper, cricket, and fly. The large green grasshopper about an inch long used to kill well in the Lee and Deep Streams.

Ten miles south of the Clutha River is Catlin's River, extending about eighteen miles inwards in a north-easterly direction, through heavy bush-country, which is very trying to the angler; but to those who can surmount difficulties ample reward is in store, as the river is full of fine fish. All kinds of baits are used here with success but perhaps the creeper is the best. Very large baskets are got here every season by the few who tackle it; it is December before the fish are at their best, except near the mouth, where they are always in good condition. Running parallel with the Catlin's, about four miles apart, is the Owake, which flows into the same bay; it also yields good baskets, and is more easily fished, the country being more open, the average for the Owake will be about 2lb., and Catlin's 3lb.; baskets of 40lb. and 50lb. are to be had in the Catlin's frequently.

The Mataura River, which is the southern boundary of the Otago society's district, will be found described with its southern tributaries under the district of the Southland society. The first tributary we come to on the Otago side is the Mimihau, running eastward amongst the hills about eleven or twelve miles, and, being free from snow-water, is in good fishing trim at the opening of the season, but the trout are not at their best until the middle of November. This stream is one of the best in the district, and a great favourite with the South anglers, good baskets being obtained during the whole season; 30lb. to 40lb. weight can be had on a suitable day, the average being about 2lb. The lower reaches next to the Mataura are mostly fished with the minnow and creeper; in the upper waters the fly is the best lure. A 7lb. trout was taken out high up this year with the fly.

To fish the lower waters, stop at the Township of Wyndham, where there are good hotels. Wyndham is the terminus of a branch line from the main south line of railway. The upper waters are more easily got at from the Mataura Bridge, on the main line, where there are two good hotels.

The Waikaka joins the Mataura about thirty-five miles up, and extends north-east for some twenty miles, also free from snow-water. The lower reaches are not so good for fishing as those higher up, although some heavy trout are taken occasionally low down. One requires to go up eight or ten miles to get a good day's fishing. The trout will average about 2lb. Gore Township, on the main line, is the best place to put up at to fish this stream.

The Waikaia joins the Mataura about fifty miles up, extending northwards for about twenty miles through a beautiful valley. It is sometimes affected by snow-water until the end of October. Being a long way from inhabited centres, it is not very much fished, but it is a lovely stream. The trout will average between 3lb. and 4lb., and are in great numbers. The few inhabitants in its neighbourhood get grand baskets every season, the fish being of a very fine quality. There are several tributaries running into it well worth attention—namely, the Dome Creek and Gow's Creek. The minnow, cricket, and fly all kill well here. Some further particulars of this river are found further on. The most suitable place to fish this stream from is Waikaia, where there is a good hotel, and is reached after a drive of fifteen miles by coach from Riversdale Station, on the Waimea line of railway, as you leave Gore for the Lakes.

The Owake is fifteen miles from Balclutha. Catlin's is four miles further on. There is railway communication to within four miles of the Owake, thence coach. Between the two streams, near the coast is a good hotel, from which both streams can be fished.

The Water of Leith is worth mentioning on account of its past history. It flows into the Dunedin Harbour through the north end of the city. It was this stream that received the first trout introduced into New Zealand, and from its waters the other streams were first stocked. It extends in a north-easterly direction for five or six miles, and is the beau ideal of a trout stream; not a yard of its waters but what is a happy home for master trout. Although quite a small stream, one could scarcely credit the quantity of trout taken out of it ever since it was open for fishing. In the early days good baskets used to be had. Mr. Russell, the honorary secretary of the Otago society, can remember seventeen years ago, when he and perhaps another solitary angler would have the whole river to themselves on a holiday. Now, you have always a dozen and a half rods in sight, and a 1lb. trout is considered a prize in its upper waters. After a fresh, trout from 3lb. to 5lb. are got in the lower reaches, having come out of the bay, and always in the pink of condition. The stream is full, of trout of small size, and many a good angler has obtained his first experience in this little stream.

The Waitati enters Blueskin Bay fifteen miles north of Dunedin. Not a very large stream, but almost as good as the Water of Leith. All the fishing season good baskets are to be had, from 15lb. to 25lb. on a favourable day; average ¾lb. There is a good hotel at Blueskin, and, by leaving Dunedin by first morning train, returning by last train, ten hours can be spent on the stream.

Thirty miles north of Dunedin the Shag River flows into the sea, extending inwards about twenty miles; being affected by snow-water it is uncertain until November, and occasionally is discoloured by diggers. Otherwise it is a fine trout river, and 50lb. baskets have been got on good days. The lower reaches contain large trout, which are taken by the minnow and creeper. The upper waters are fished principally with the fly and creeper; at times good baskets are to be had with the worm, this being about the only stream on which the worm-fisher is successful. On the lower reaches the trout will average from 3lb. to 4lb., higher up about 2lb. Railway communication and hotel accommodation every few miles along its banks for twelve or thirteen miles, Dumbeck being the terminus, which is eight miles higher up than Palmerston on the main line.

There are numerous other small streams in Otago not mentioned here, but well worth fishing; and the society are rearing and distributing the Scotch burn trout into all the hill-streams, in which the larger kinds of trout do not care to remain long.

The Southland District.

As has already been stated with regard to other parts of the colony, most of the streams in Southland are practically free to anglers. No difficulty will be experienced by the fair fisherman (or woman), having a license, roving where he will, either as regards fishing or camping. As a general rule, on the larger rivers there is a chain reserve on the banks, and the rivers are thus entirely open to the public; and, as regards the few exceptions, he will not be interfered with if he avoids doing any damage, in breaking down fences, leaving gates open, driving sheep, or walking through standing crops, in fact, respecting other people's property as he would his own. One other piece of advice: an angler should never have a dog with him, both for his own comfort and that of other people.

Of all the rivers, exclusively confined to the Southland District, the Oreti River takes the palm, being the most prolific of trout owing to its superiority as a trout-breeding stream. This arises from its suitable spawning grounds, from below

Dipton here and there upwards towards its source in the Eyre Mountains. It is a shingle, snow-fed river, open and accessible all the way, with a public reserve on each side, with wide banks only reached in floods, and often changing its course from one bank to the other, with many turns and variations in depth, with swift runs, ripples, and eddies at the heads and tails of a succession of comparatively still pools. A number of lagoons, mostly of a crescent or horseshoe shape, are met with on both sides of the river, which are useful nurseries to the trout, and make roving along the banks difficult without wading. Wading, however, is not absolutely necessary, although double the quantity of fish will be caught by the wader.

From the opening of the season down to about Christmas, the Oreti and the New River (the name given to the Oreti below Wallacetown) are subject to be occasionally discoloured by snow-water. This, however, only prevents fly-fishing, and the latter is at all times of very little good below Benmore. The minnow, alive, dead, or artificial, is the bait; the creeper, which does not last beyond Christmas, is not of much use, and besides is practicably unobtainable about this river. The whitebait phantom is most used, and with this, taking the whole river, the largest bags are made.

From Christmas, however, when the water always clears and soon becomes very bright, the fly, both artificial and natural, takes the best above Benmore, except at night, when the minnow is still used. This is especially the case when the cricket, or, more properly, the cicada, becomes plentiful, which is soon after Christmas, and lasts till the middle of March. Of all natural baits used in fly-fishing the cricket, a beautiful large insect with large gauzy wings, takes the cake, or rather the trout, and with a little practice it can be thrown with a long two-handed rod as far as the artificial fly. It is found most plentifully in the short silver tussock and cutting grass, abounding near the river, and can be used with success on the Oreti all the way above Winton and for a few miles below.

It seems surprising that trout can be caught with spinning the natural minnow, or, for preference, the smelt or whitebait phantom, when the water is much discoloured. The honorary secretary of the Southland society informs me, however, that he has caught them up to 9lb. in weight, spinning on the shallows when the water is in this condition.

When the water has cleared in the lower reaches where fly-fishing ceases, that is, a little below Winton, not much can be done with the minnow on bright days, unless there is a stiff breeze up-stream from towards the south, or a burster from the north-west, except at night; but, be it ever so bright, if there are good ripples from a strong wind you may have sport.

Some of the largest bags have been made in the roughest of weather—wind and rain—when one can hardly keep the bait on the water. The trout vary in size from 1lb. to 12lb., or even more, according to locality, the higher you go the smaller and the lower, towards the New River Estuary, the larger. From Wallacetown up to Dipton you may put the average at 3lb. or 4lb. They are very capricious in the days and hours of feeding, and they exhibit the usual decidedly migratory instincts. Many are caught by the fishermen in the New River Estuary, and they have travelled thence through the New River heads around the coast to the Bluff Harbour, a distance of some twenty miles in the salt sea, where large specimens of the brown trout have been seen swimming about the piles of the wharf. In the New River, the name given to the Oreti south of Wallacetown, the trout resemble in appearance the white trout of Ireland. They have no red spots, but large brownish X-shaped ones. Enormous numbers of trout are taken out of the Oreti and New River in the season by the numerous anglers.

Centre Hill is about the highest point accessible on the river where there is an hotel. This is about twenty miles west of Lumsden, which is on the Invercargill and Kingston line of railway, and can be reached from Lumsden by coach or trap. The next place, coming down the stream, is Mossburn, where also there is an hotel, to which a branch railway from Lumsden extends, but there is only a weekly train, the distance being ten miles. From Lumsden, down to two or three stations below Winton, the river can be easily reached by walking from any one of the numerous stations on the Invercargill and Kingston line (mentioned in the railway time-tables, which give also the distances in miles).

At Centre Hill the first tributary of any importance joins the river. This is Centre Hill Creek, where there are numerous trout though smaller than in the main stream, and where wading is not necessary. A little above Lumsden, the Irthing, the chief of the small streams called the Five Rivers, joins the Oreti. At Dipton the Dipton Creek joins, but it is small, and the trout do not exceed a pound in weight. A similar creek called the Winton Creek joins the river at Winton.

The only other, and the main tributary, is the Makarewa itself, a large river, which joins the Oreti a little below and some distance to the west of Wallacetown Junction, where the Makarewa is crossed by the railway. From here to the north, as stated above, the Oreti, thus materially increased in size and volume, is called the New River. The chief stopping-places on the Oreti below Lumsden are Dipton, Benmore, and Winton, where there are comfortable hotels.

Lumsden makes a very good centre, both for the Oreti and

the streams known as the Five Rivers, of which the Irthing is the principal. These flow into the river between Mossburn and Lumsden, and can be fished either from the Lumsden, Mossburn, or Lowther Railway-stations. In these there is good fishing to be had throughout the season, but fish become more difficult to take in February and March. Fly and cricket are preferred, and, next, live bait. All these streams are practically free, the owners and occupiers of land at Five Rivers never refusing permission of visitors to fish. They are mostly shingle-streams, with little or no bush. Occasionally, but not often, is wading necessary, and the trout run from 1lb. to 5lb. These streams are not much fished, on account of the abundance of trout in the Oreti, into which they flow.

Dipton is a township on the Oreti. Here many parts of the river are accessible, there being a series of beautiful pools and ripples, high banks on one side alternating with shingle beds on the other, making wading necessary. Some of the takes in this part of the river during last season are worth mentioning, one rod obtaining 338 fish, weighing 1,117lb., and another 560 fish, weighing 672lb. Some of the largest daily takes with minnow were: thirteen fish, weighing 51lb.; fifteen, weighing 42lb.; twenty weighing 60¾lb.; and, with fly, thirty-eight, weighing 36lb.; twenty, weighing 34lb., forty-two, 46lb.; and fifteen, 32lb.

For the Benmore district, Harrington's Crossing is the nearest railway-station, a mile and a half from Benmore Hotel, which is situated on the banks of the river. The river here is entirely open to the public, and affords famous sport, trout being plentiful and large, taking almost any bait, the phantom whitebait being the favourite spinner, some heavy fish being taken with it in the season of 1891–92. With the cricket, too, some good takes were obtained during the summer months, and also with the artificial fly, though those taken with the artificial fly here, as elsewhere, are, as a rule, not so large as those caught with the minnow or cricket. The banks of the river are low, without obstruction, but little can be done without wading.

Mention should here be made of the Otapiri River. This is distant six miles from Centre Bush Station, eight miles from Benmore Siding, and about the same distance from Harrington's Crossing. It is also open to the public, and can be reached either in a trap or on horseback; at any part of it there is good camping-ground, and a plentiful supply of firewood.

The Otapiri is a beautiful mountain stream, the hills in many places coming to the water's edge; the upper part of this stream for several miles has a very rough bottom, the water running rapidly over rocks, and, in some places, large boulders, until the level country is reached, then it becomes deep and sluggish. Good sport can be had with trout up to 5lb.: they take the fly readily, and are very strong and lively, the best fishing being down to its junction with the Lora. The banks are low, with a little flax occasionally, and scarcely any wading is required: the best time is in the early season. As no snow comes into this stream, it gets low in dry weather. Mr. Hunter, in the season of 1891-92, caught here several nice baskets from twelve up to two dozen in a few hours with the artificial fly.

Winton is nineteen miles north of Invereargill. The Oreti, south of Winton Bridge, is seldom if ever fished otherwise than with minnow. The whitebait soleskin is generally used, and most anglers prefer the smallest size (about 1½in.). The natural minnow and smelt are plentiful, until a month or six weeks before the season closes, when they begin to disappear. These make a splendid bait, either spinning or used alive. They are not generally used however, on account of the delay and trouble in obtaining them, and their being somewhat troublesome in use. This river is subject to floods, and, flowing generally through shingle, the creeper is not to be met with except at rare intervals. This bait, so deadly where it abounds, is not so attractive to the trout in the Oreti on account of its scarcity. Here and there the sportsman's course is impeded for a short distance in some parts of this river by a lagoon or back-water, making it necessary to diverge from the river-bank, when all that he need care to do is to close such gates as he may open, avoid the immediate precincts of a farm-house, and take no dog among the farmers' sheep.

The Winton Bridge is about twenty miles from Invereargill, and two from Winton, which is on the Invercargill and Kingston railway line, and the bridge is reached by a good road. The evening and early morning are the best times for fishing the Southland rivers generally, and the Oreti is no exception. A dull day and a good breeze are, however, pretty certain to insure good sport throughout the day if the river is free from flood. Should it be swollen, it is best to sink the minnow well below the bank, and draw slowly but spasmodically upstream, when the fish lying underneath the bank will often rush at it, especially if the flood be not very high.

There are nearer points where the Oreti can be fished. Oporo, which is a small station on the Riverton line beyond Branxholme, is only about twelve miles from Invercargill, and can be reached either by rail (see railway time-table) or by trap, as there is a good road. Another point easily reached by trap from Invercargill is the Iron Bridge, distant nine miles on the same road, a little beyond Wallacetown, where one can fish either above or below the bridge, though the latter fishing seems preferred. The fish are not quite so numerous as at and above Winton Bridge; but they are generally larger, running from 2lb. to 15lb. in weight, with an average probably of 4lb. to 5lb. At a distance of about four miles below the Iron Bridge, after the Makarewa joins, the water is tidal and sluggish, so that the artificial minnow is comparatively useless, and the natural minnow or bully should be used either living or spinning. This applies particularly at the New River Ferry Bridge, which goes across the New River about two miles below the junction of the Makarewa, and is distant only seven miles from Invercargill by a good road. This bridge is distant six miles from the mouth of the New River in the New River Estuary, and all the way down there is good fishing in the early part of the season. The best plan is to take a trap for the day, with luncheon, and a billy for making tea; if you have good sport the trap will be required to bring home the catch.

On one such occasion early last season, one rod, fishing from noon till 4 p.m., inclusive of half an hour for luncheon, caught with live bully four fish, weighing 12lb., 12lb., 10lb. and 9lb., or 43lb. altogether. Probably in consequence of the great abundance of minnow life in these lower reaches, and of the fact of the flow of the tide bringing up smelts and whitebait, the trout are capricious and difficult to please, and, though dashing about and breaking the surface just under one's rod, they generally object, later on in the season, to the size or appearance of the live or spinning natural bait, and exasperatingly prefer the smaller fry, such as especially the whitebait. The largest catch that has been reported in the neighbourhood of the Iron Bridge in one day was nineteen trout, all caught with the whitebait phantom. These averaged nearly 4lb.

The Makarewa, although a tributary, deserves notice as a distinct river on account of its length and size and different characteristics. Here the trout in the lower reaches are very fine in quality, and distinct in appearance from the New River trout. They have red spots, and are reddish in appearance, instead of a light-brown like the New River and Lower Oreti trout. The banks of this river are confined, and the bed alternately earthy and gravelly, with very deep holes and pools. About Wallacetown and the lower reaches the banks are so high, and beautifully wooded here and there, that to rove with a fly or minnow would require a balloon. It is only here and there you can get at the water at all, and the fish seem to feed exclusively at night, good baskets having been made only then, the trout running up to 14lb. Two or three miles up the river the banks soon decrease in height, allowing access to the stream without difficulty. This fishing is easily accessible from Invercargill. Wading is not necessary, and is not of much use.

FIG. 1
SHAG RIVER—Male 4½ lbs., length 20 7/10 ins.

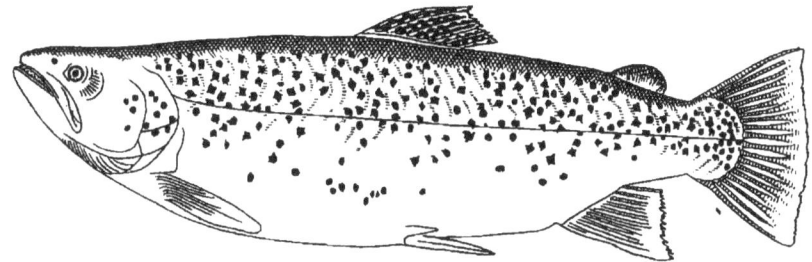

FIG. 2
WAIKOUAITI RIVER—Female 13 lbs., 11 oz., length 28½ ins.

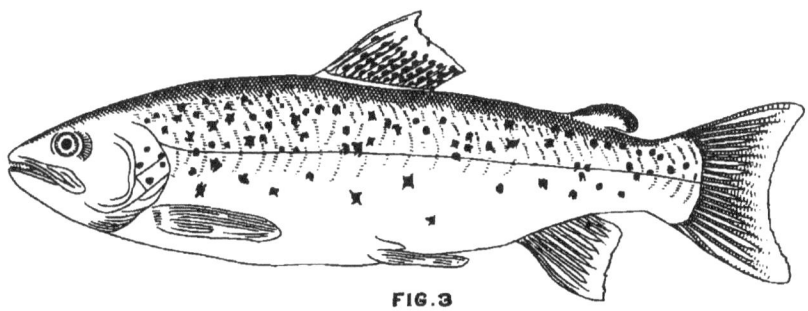

FIG. 3
LEE STREAM—1 lb. 6 oz., length 13 1/3 ins.

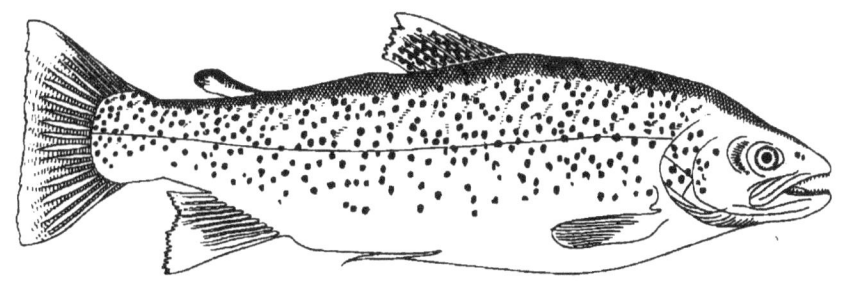

FIG. 4
WAIWERA—Female 2 lbs. 13 oz., length 18 ins.

OTAGO TROUT (*S. fario Ausonii*) from different Rivers.

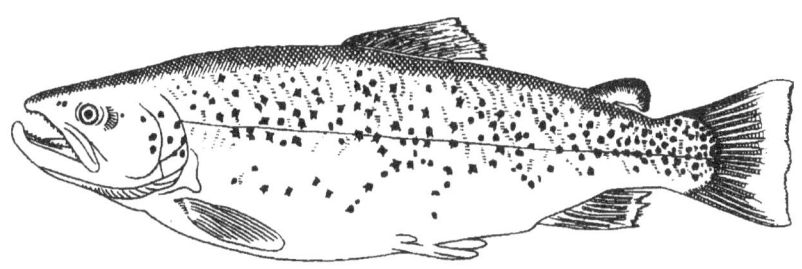

FIG. 5
WAIPAHI — Male 8 lbs. 10 oz., length 24 3/4 ins.

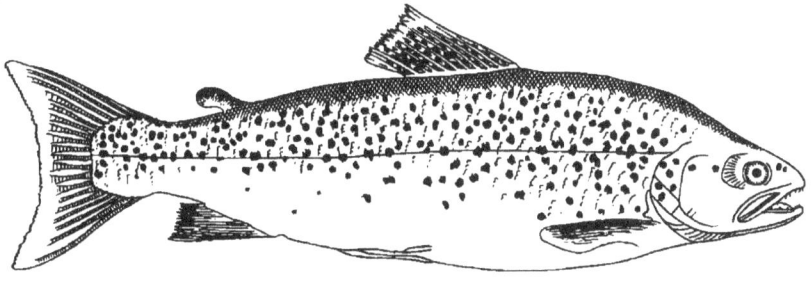

FIG. 6
WAIPAHI — Female 2 lbs., length 16 inches.

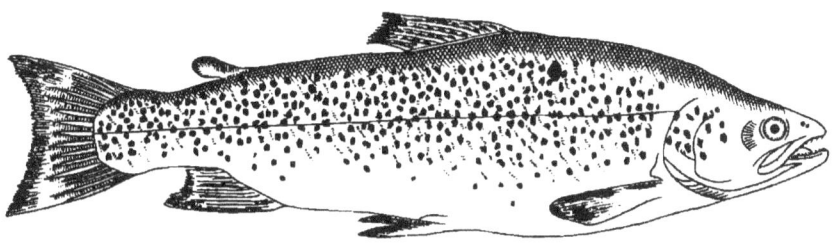

FIG. 7
POMAHAKA — Female 3 lbs. 2 oz., length 19 1/2 ins.

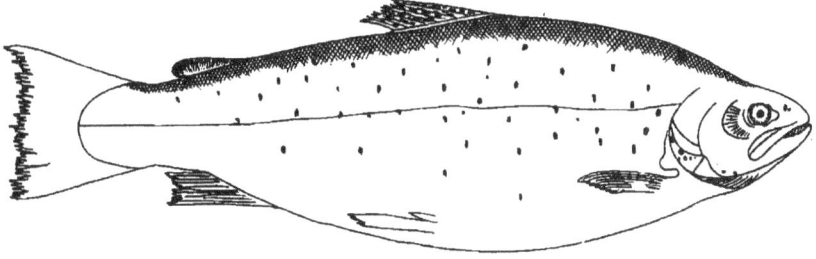

FIG. 8
CANTERBURY TROUT — Female 2 lbs., length 15 6/10 inches.

OTAGO TROUT (*S. fario Ausonii*) from different Rivers

About twenty miles up, at the head of the Makarewa, it branches off into the Hedgehope on the right, and the Lora and Otapiri—referred to previously—on the left. In the Hedgehope there are plenty of trout from 1lb. to 3lb., as also in the Dunsdale Creek, a small stream running into the Hedgehope ; but it it is twenty-five miles or more from Invercargill, and camping would be necessary.

The mouth of the Makarewa, and a few miles of many windings towards Wallacetown are approachable only with a boat, as the banks are forest-lined, making a beautiful avenue of water through the bush. This is the only part obstructed by bush to any extent of all the Southland rivers, except, however, the Purakino River, which runs into the Aparima Estuary at Riverton, and is lined with bush, part of the Longwood Forest. For the lower waters of the Makarewa use the minnow, and the upper reaches the fly, and especially the cricket in its season, or the creeper, which comes in before the cricket.

The Oreti and its tributaries form the centre of the district. Taking the streams to the west of the Oreti, the first of any importance is the Waimatuku, a small stream, across the whole breadth of which the fly can easily be thrown. It is nine miles from Riverton, and fifteen miles from Invercargill, and can be reached either by road or rail. Running through flat country it is a sluggish stream. October and November are the favourite months, there being more water and absence of weeds, which, when the river gets low in the summer months, grow up from the bottom. Both live bait and fly are used, and trout have been taken up to 15lb., and, since the Southland society have stocked it regularly, fish are more plentiful, but not so large. The trout here are of a dark colour, on account of the swampy water, but of very fine flavour. Wading is desirable to reach the pools, the banks alternating with shingle and flax.

Next to the west comes the Aparima, known also as Jacob's River. It has its rise in the Takatimo Mountains, and after a course of about seventy miles discharges its waters into the ocean at Riverton, where it forms a large estuary, a great portion of which is left dry at low tide. This river was originally set apart by the New Zealand Government to be solely stocked with salmon, and consequently no trout were put into it by the Southland society. Unfortunately, however, several landowners in the district, becoming impatient at the failure of the early attempts to import salmon ova in the early days, when direct communication with Great Britain was only with sailing vessels, obtained trout fry from one source or another, and with them stocked several tributaries of the Aparima. The natural consequence of this ill-advised action is that the main stream itself is full of trout, and the difficulty of deciding the question of the acclimatisation of salmon is increased. The fact that brown trout in New Zealand show such a strong liking for salt water, and the difference in appearance and in other particulars which this living in salt water has given them, has made their identification a puzzle to the best experts.

It may be of some interest to refer to the different opinions that have been given with reference to some fish taken from the Aparima. As before stated, very many thousands of salmon-fry have been turned into this river. In May, 1891, the estuary was netted and several fish taken: these were distinctly different to any New Zealand form of brown trout. Some were sent to Professor Hutton, one of the best authorities in the Southern Hemisphere; and to myself; and some were sent to Dr. Günther, Curator of the Natural History Department of the British Museum. I obtained, for the sake of comparison, a trout from the Acclimatisation Gardens at Christchurch that had always lived in confinement, and also a specimen (*feræ naturæ*) from the Selwyn River, both of about the same weight as the specimen sent from the Aparima. From an examination of the number of the branchiostegals or gill-rays, the form of the operculum or gill-cover, the number of the fin-rays; the position, size, and shape of the teeth generally, and especially on the vomer; the size and form of the scales generally, and the number from the root of the adipose fin to the lateral line specially; the form and position of the spots on the gill-cover; the length of the maxillary, and the number of the pyloric cœca, I concluded that the fish were grilse, and Professor Hutton in the main concurred in this conclusion. Dr. Günther, however, in his account of those forwarded to him, says (see *Field* newspaper, 9th January, 1892): "The specimens are most assuredly not salmon (*S. salar*); neither are they brown trout (*S. fario*). They are a kind of sea trout (*S. trutta*), looking extremely like the Irish white trout. But the different kinds of migratory sea trout are so closely allied to each other that it is almost a matter of impossibility to give an opinion on artificially-reared fish or their offspring."

Now, the extraordinary peculiarity of this opinion is that Dr. Günther says these fish were certainly not salmon (*S. salar*) nor *S. fario*, these being the only two kinds of fish that have ever been put in the river, while he calls them a kind of sea-trout (*S. trutta*), which have never been placed either in this river or in any other within scores of miles of it. In other words, he says they were certainly neither of the kinds of fish that were placed in the river in thousands, but another kind that has never been put there at all.

But, whatever they are, there are large numbers of salmonidæ in the Aparima Estuary from 1lb. to 12lb. in weight, and Mr. A. N. Campbell, the curator of the Southland society, informs me that he has often seen them caught in the same net at different times with red-cod (*Lotella bacchus*), barracouta (*Thyrsites atun*), dogfish, elephant-fish, mullet (not the *Mugil*, but *Agonostoma forsteri*), flounders (*Rhombosolea monopus*), &c., a plain proof that they can live and thrive with these voracious fishes as soon as they have attained a certain size.

The number of salmon-fry turned into the Aparima by the Southland society during the last few years is as follows: In 1886, 5,500; in the latter part of 1887 and 1888, 52,000; and in 1890, 87,000; and the Otago society have liberated in the same river a very much larger number.

The prohibition by the Government of all angling for trout in this river has caused some little dissatisfaction amongst the settlers, and representations have been made to have the river declared open. The importance of the salmon experiment is, however, sufficient justification for keeping it closed for the present. After a time it will probably be opened, and therefore some account of it will be useful.

The Aparima, in its long course through the fertile valley which bears its name, presents every variety of water which the angler loves and the trout or salmon delight to revel in. Part of its upper waters is of a rocky character—heavy streams with deep pools and a rough boulder bottom predominating. Again it flows between grassy banks, with heavy streams and deep reaches alternating. Although there are plenty of open beaches in its course it is by no means a shingle-bed river, nor is it affected by snow-water to any great extent. Altogether, the Aparima is an ideal stream from an angler's point of view, producing short and thick fish of the finest quality.

That part of the river under tidal influence could be fished from Riverton, and no doubt good bags of large trout would be made there with minnow or live bait. A few miles further up is Thornbury Junction, with a good hotel near the river. Seven miles further is Fairfax, and five miles on is Otautau; both have good hotels and are close to the river.

The Riverton-Nightcaps line stops at each of these places, and runs daily. Twelve miles up the river from Otautau there is a hotel, at Wrey's Bush; but to reach it one must drive from Otautau, where good conveyances can be got from the hotelkeeper. This covers over thirty miles of excellent water, but to fish the upper waters the angler must camp out, the country being very suitable for this. Fly-fishing should be very good on the higher parts, and minnow most effective on the lower portion of the river.

The Otautau.—This is a tributary flowing into the Aparima. The Otautau runs through the township of that name, distant from Invercargill thirty miles. From Invercargill there is a daily train.

The best months are November, December, January, and February. The minnow (artificial and natural) seems to be the best lure in this stream, although occasionally fish are taken with both fly and cricket. The river has a half-chain reserve on each side, and is consequently public. The upper portion has confined banks, and is rather sluggish, but the lower part is shingle, with fine ripples and streams; for the latter portion waders are necessary. The general sizes of trout caught would average $2\frac{1}{2}$lb. to 3lb.; fish have been taken up to $15\frac{1}{2}$lb., and it is quite common to take six to eight fish of the average size in the course of an afternoon. There is plenty of hotel-accommodation for anglers in close proximity to the river, also plenty of camping-ground for those so disposed. The best time of day is generally the evening, but this, like many other streams, depends greatly on the state of the water and the weather.

Again, west of the Aparima are the Morley and Orawia. The nearest fishable water in the Morley is five miles from Nightcaps, which is situated fourty-four miles north-west of Invercargill, and is the terminus of the Riverton-Nightcaps Railway, and contains an hotel, post- and telegraph-office, and the usual stores, &c., common to all country townships. The daily train on most days arrives from town in the evening, and leaves for town in the morning, so that tourists can get to or from Invercargill, Queenstown, or Dunedin the same day. A horse can be had, and occasionally a trap, by giving notice beforehand, and there is a good summer road all the way; and, as there are a few flax-bushes near the stream here and there, the horses could be tethered while fishing, and there is ample rough feed at hand, though it would be as well to take some.

The Morley—a rain-fed stream—rises in the Beaumont Hill, and receives one or two tributaries before it is joined by the Orawia near the Birchwood and Mount Linton stations, ten miles from Nightcaps. It is here that perhaps the best trout-fishing is to be obtained, although it is also good all the way down the river, from the junction, called the Orawia, until it finds its outlet in the famous Waiau River. The lower part of the Orawia may also be reached from Otautau Railway-station (thirty miles from Invercargill), in the same line, where traps and horses can always be had to the Eastern Bush Accommodation-house, a distance of about twenty miles, and not far from the river. A great part of this road is through the Merrivale Estate, and near the river are the wonderful limestone caves, which themselves are worth a visit—many visitors coming every summer to explore their various and extensive caverns.

The Morley and the Orawia, below the junction, are streams that at the first glance delight the heart of the angler; gravelly open beaches, with little or no flax or other obstructions to casting, the bed of the river for the most part consisting of a soft blue clay-stone, almost like rock in the distance, which accounts for the very clear water, except when it is in flood. There are nice ripples and curly pools below them every hundred yards or so, with occasional fairly-long stretches of dead-water between; in the latter it is not much use fishing unless there is a strong breeze on, when the fly, blowfly, or cricket may be used with success.

The grub, or creeper as it is here called, is found below the stones near the margin of the stream up to Christmas and even later, and sometimes proves a deadly bait; but this river may fairly be called the "phantom" one, for the trout take it readily even during the day when the wind is favourable—and it may here be said, in passing, that, as a rule, the east wind is generally unfavourable for sport; but as there is not much of it in Southland during the season the angler is rarely disappointed in this way. Bullies and other live bait are to be had for the catching, and will probably prove more deadly than the artificial; but, of course, this entails extra trouble. Flies will take during the early part of the season, the best being the March brown, red and black hackle, and governor—which, indeed, may be called the standard flies for the Southland District. There is no reserve on either side of the river, which, for the most part, runs through private property, the owners of which have never been known to refuse access to any true sportsman. It is fordable in ordinary weather. Waders are advisable, otherwise many choice casts may be missed. Few small trout are caught, the usual size being from 4lb. to 8lb., the average being about 5lb., and five or six a day may be considered fair sport. The minnow-tackle should be strong, but not too stout, owing to the clearness of the water, and it is needless to say that up-stream fishing is the best, and keeping out of sight if possible—almost indispensable if the angler wishes to secure fish—when the water is clear. The river has been annually stocked with young trout of different kinds for some years past, and as it is not much fished it ought, in a few years more, to yield better sport than hitherto; but it is not probable that the trout will be heavier except in the lower water, which experience has shown the larger fish make for in time.

Just stopping to mention the Wairaki (which falls into the Waiau twelve miles beyond Mount Linton), with its beautiful clear blue water, with shingle and pebbles, and some fine trout in it, as being at present too far out of the way of the angler, we pass on to the Waiau River—a mighty stream, draining the great Te Anau and Manapouri Lakes—very large, deep, and rapid. On account of its great size, depth, and swiftness, and its comparative remoteness from settlement, it is unfished, and practically as yet unfishable to any profitable extent, though there are trout in it of enormous size. It falls into the sea six miles above Orepuki, the terminus of the Orepuki Branch-railway from Riverton.

The only other tributary of the Waiau on the Invercargill side of it is the Mararoa, falling into the Waiau below Lake Manapouri. This river drains the Mavora Lake, far to the north, and can be easily reached by the coach running in the summer from Lumsden to Lake Te Anau. You pass beyond Centre Hill, and are dropped at a new hotel *en route*, which is situated near the river at its best part. The river is swift, running over a rough bed of shingle, boulders, pebbles, and gravel, with clear and low banks. Wading, of course, is preferable, though not absolutely necessary. The fish run large, a ten-pounder being frequently met with. Minnow-spinning, dead or artificial, is the mode employed, excepting in the cricket season, when the cricket takes best; and the scenery, being near to the Takatimos and Mount Hamilton, is very fine.

Having now exhausted the principal streams to the west of the Oreti and its tributaries, we turn to the east.

On this side the nearest stream to Invercargill is the Waihopai. This small river rises about Morton Mains, on the Invercargill and Dunedin Railway, and falls into the New River Estuary, there called the Waihopai estuary, two miles on the north road from Invercargill. It is much obstructed by bush towards its mouth and by flax here and there further up. The best fishing is at One-tree Point Railway-station, on the east road, six miles off. It has been heavily stocked with fry by the Southland society, but it runs very low in a dry summer, and the trout are large but not very numerous. When it is blowing half a gale is the best time in the day, as the water is clear and rather peaty. No wading is required. The fish are of good quality, and go from 3lb. to 10lb. or more in weight, the minnow only being used—the natural for preference. The angler must be content if he takes two or three of these big fish in a day. Just at and after dark they generally take freely.

There is no other stream on this side until you come to the Mataura River, at the Mataura Township, on the same line of railway, and which river forms the eastern boundary of the Southland society's district to the sea. This is a magnificent river, longer and larger than the Oreti, which it resembles in some respects, with less shingle and more gravel and sand. It has one disadvantage in being very often discoloured at other times than during floods, owing to the discolouration through mining operations which takes place in the Waikaia River, flowing into the Mataura at the Pyramids Bridge, opposite Riversdale, on the Waimea Plains.

There is no tributary to the Mataura River on the Invercargill side from Mataura Town all the way down to the sea, the tributaries in this distance being on the other side of the river, and therefore included in the Otago acclimatisation district—viz., the Mimihau and the Wyndham rivers, which have their mouths near Wyndham Town. We turn, therefore, upstream from Mataura, and the first stream on the west side is the Waimumu, between Mataura and Gore. This is very small, but the trout thrive, and are numerous, it being continually stocked by the Southland society. Then we come to the Otamita, running into the Mataura at Mandeville, on the Waimea Plains Railway, distant from Gore about ten miles. It has its source in the Hokanui Hills. It is entirely dependent on the rainfall for its supply of water, consequently in a dry summer it gets so low as to be almost unfishable, and the trout become an easy prey to the poacher. For the first three months of the fishing-season, however, this is, without exception, the best fly-fishing river in Otago or Southland. A few years ago the trout averaged over 4lb. in weight; now they run smaller. Mr. A. N. Campbell fished this river off and on for three months one year, and caught 120 trout, which averaged a little over 4lb., the heaviest being 7½lb. All of them were caught with fly, and he generally gave it up when he thought he had over 20lb. in the bag. Three seasons ago—1889-90—the summer was very dry, of which the poachers took full advantage, and almost cleared the river of large trout; also, the local anglers found out what a deadly bait the creeper is, and with it made enormous bags. As much as 70lb. were taken by one rod in a day, and baskets of from 40lb. to 50lb. were very common. This is altogether too much for a small river like the Otamita if the stock of large trout is to be kept up. The trout in this river now average about 2lb., and this last season—1891-92—there was a marked increase in weight over the previous one, so that, if the river only gets fair play, the fish will soon recover their former average.

There are fully twelve miles of good fishing-water with clear open well-defined grassy banks and no shingle-beds. Good accommodation can be had at the Railway Hotel at Mandeville, close to the river on the Waimea Plains line, which connects with the express from Dunedin to Invercargill at Gore every day. Most of the Otamita runs through private property, but the proprietor very kindly makes all true anglers welcome. A fairly skilful angler ought to get from 20lb. to 30lb. on a good day with fly in the beginning of the season, or with cricket later on. The flies found most successful are

March brown, black gnat, red quill, and otamita, which latter is a fly Mr. Campbell discovered some years ago, and which is now sold under that name at all the fishing-tackle shops.

The only other tributary to the Mataura River on the Southland side of any importance is the Waimea Stream, which is nearly the same size and length as the Otamita, and falls into the Mataura a little above the stream. The lower reaches of these two streams are very near together, so that an angler can fish both of them with ease. While the Otamita is essentially a stream for the artificial fly, the Waimea is rather for the minnow and the cricket, and is the less attractive and prolific, being a somewhat sluggish stream.

Mention has been made of the Waikaia, which flows into the Mataura on the Otago side of the river some twenty miles above Gore, and of the discolouration caused to the Mataura below the mouth of the Waikaia from the washings from the gold-diggings. Yet this part of the Mataura is full of trout, and large numbers are caught every season with minnow, and also with fly when the water clears through the discontinuance of mining operations. During the season 1891–92 Mr Sparks, of Gore, took on one day from the Mataura, with fly, fourteen fish, averaging 3lb., and also had some excellent takes with minnow and Archer spinner. Mr. Cumberbeach, of Gore, caught, in three hours' fishing one afternoon with cricket, eleven fish, weighing 40¾lb., and on another afternoon nine fish with artificial minnow, weighing 37lb., the heaviest being 5½lb. These are some of the best takes; but it is rare that any expert angler fails to get a good basket of trout when the river is in anything like order. Above the Pyramids Bridge, on the Waimea Plains, the Mataura is clear after the snows have melted on the ranges, like the Oreti. This is about Christmas, and thereafter there is first-rate sport to be had all the way up to Garston, on the Kingston line of railway, and a little beyond. The river runs parallel to Riversdale and Longridge, on the Waimea line, but on the other side of the plains under the hills, and distant from three to seven miles. Opposite Longridge, however, is the outlet of the Nokomai Rocky Gorge, through which the river flows, and this soon stops the angler. He must then take the rail through Lumsden and Lowther to Parrawa and Athol. There is an accommodation-house at Parrawa, which is at the head of the long gorge; also at Athol, a more central point, whence the angler can fish unobstructed either up the river by the Nokomai Station to Garston or down to Parrawa. By taking a trap from Athol he can spend a delightful day in the gorge, crossing the river three or four times, and going down the gorge some miles. During the months of January, February, and March there is nothing to beat the cricket, which is here abundant; but when the water is discoloured by rain good sport can be had with the minnow, spinning the natural bait for preference. When, however, the water should happen to clear earlier in the season there is first-rate sport with the artificial fly, such as the March brown and governor.

Here must conclude the description of the Southland district, and with it this short account of the trout-fishing to be had in New Zealand. Many rivers have been passed over with a short and curt description; many of the smaller ones have not been mentioned at all; yet sufficient, I hope, has been written to give inquirers some outline of what sport is to be had. In Mr. Seton-Karr's book ("Ten Years' Travel and Sport in Foreign Lands," pp. 238-239), he gives an account of some trout-fishing in south-east Finland, where the weights of the fish caught appear to be equal to ours; but the water is the property of a wealthy and exclusive club, and apparently very limited in comparison with the enormous range of perfectly free trout-fishing afforded to the angling fraternity here.

Lastly, I should like to add, for the satisfaction of any English reader, that nothing has been inserted in this little work on hearsay, but all the numbers of trout and weights mentioned have been certified as correct either by some acclimatisation society or well-known colonists. If there is any part of the world where trout are larger and more numerous than in New Zealand, nobody will be more pleased to hear of it than myself; but till then I think we may fairly claim, in this record-making era, to have established one both for weight and numbers of *Salmo fario*. It is not too much to say that hitherto no New Zealand angler has dared to speak the truth at Home about our fishing for fear of being looked upon as an unmitigated "something," and many a colonist can look back with pain on the polite silence with which his perfectly truthful statements have been met in many a London club. If this little work tends to remove their suspicions of our veracity, to assist the many thousand anglers in England to a knowledge of the fishing to be had in this little colony, and to tempt only a few of them to come and enjoy it for themselves, I shall feel that it has not been written in vain. Begun originally as a mere pastime to fill up a few spare winter evenings, it has ended by taking up most of my leisure time for several months; yet I can scarcely look back upon this time as misspent if the information here afforded in any way assists hereafter a few of my brethren of the rod to a tithe of the pleasure I have derived during the last ten years from an occasional day on some New Zealand stream.

Christchurch, New Zealand, September, 1892.

APPENDIX.

List of Requirements for any Rain-River.

A 12ft. fly-rod, check-reel and 40 yards fine line, creel or bag, 3-yard fly-casts, flies, silk and wax, No. 6 or No. 7 plain hooks (eyed upwards), shot, or small coil of fine lead wire; small tin box for carrying creepers or beetles; small round tin box, or equivalent, with sponge or flannel for keeping moist some spare flies and casts; disgorger, landing-net or gaff, pocket-knife or pair of folding scissors, bait-can if live bait to be used, small net for bait-catching.

List of Requirements for any Snow-river.

A 14ft. or 15ft. rod, check-reel and 60 yards line, strong, but not too thick or heavy; creel or bag, traces with at least two swivels, 3-yard strong gut casts, tin box containing various spinning-baits, such as the whitebait phantom, Devon, and others; silk and wax, No. 6 or No. 7 plain hooks (eyed upwards), lead wire, some strong spare gut for repairing the mountings of spinning-baits, spare swivels and triangles, disgorger, gaff, small round tin box, or its equivalent, with wet sponge or flannel for spare trace; pocket-knife or pair of folding scissors, bait-can, bait-net.

Camping Outfit.

A tent for two, three, or even four persons, 8ft. by 10ft., with fly at least 10ft. by 12ft., or, better, 12ft. by 14ft.; ridge- and tent-poles; twenty-four iron pins of ¼in. iron 10in. long, for ropes; some smaller pins or wire bent thus ∩ for fastening sides of tent; two billies, each holding at least two quarts; plates, frying-pan, coffee-pot, small kettle, candles and candlestick, with 3 yards copper wire to support candlestick from ridge-pole; small tomahawk, file, cups, knives, forks, spoons, tin-opener, provisions, blankets; one air-bed, with pillow and bellows, 80in. by 36in. If no air-bed, by all means have plain waterproof sheet, 6ft. 6in. by 3ft. 6in.

Regulations for Trout-fishing, Eastern Acclimatisation District, South Island.

1. Licenses to fish for or take trout in all the waters of the South Island Eastern Acclimatisation District (with the exception of the Aparima River and its affluents and confluents, in which fishing is prohibited) will be issued by either of the secretaries of the North Canterbury, Ashburton, South Canterbury, Geraldine, Waitaki-Waimate, Otago, Lakes District, and Southland Acclimatisation Societies, and for every such license a fee of twenty shillings will be charged : Provided that it shall not be obligatory upon any of the said secretaries to issue a license : Provided further that it shall be lawful for any of the said secretaries to issue licenses for the whole season to ladies, or to boys under the age of sixteen years, for the sum of five shillings each ; and to youths between

the ages of sixteen and eighteen years of age for ten shillings each ; and to men, on and after the twentieth day of December in any year, for the sum of twelve shillings and sixpence each.

2. Licenses when issued as aforesaid for the whole season shall entitle the person named therein to fish for or take trout in any of the said waters from the first day of October in any one year to the fifteenth day of April in the year following; but no such license shall confer any right of entry upon the land of any person without his consent.

3. No person shall fish for, take, catch, or kill, or have in his possession' or attempt to fish for, take, catch, or kill, in any manner whatever, any of the salmonidæ or trout, except during the above-mentioned period.

4. Every such license shall entitle the person named therein to fish for or take trout with one rod and line only, and with the following baits : Natural or artificial fly, natural or artificial minnow, and any small indigenous fish, grasshoppers, spiders, caterpillars, creepers, and worms.

5. No person shall use any other bait, or any method, device, or contrivance of any sort or kind whatever, for the purpose of fishing for, taking, catching, or killing trout, except a rod and line, and a landing-net or gaff for fish taken with rod and line.

6. No cross-line fishing, stroke-hauling, or any other unsportsmanlike device shall be used for the purpose of taking, catching, or killing trout; nor shall any of the baits above mentioned be used with any medicated or chemical preparation whatever.

7. No person shall fish for or take trout without a license, and every person fishing for or taking trout shall, on demand of any ranger, constable, or person producing a license, produce and show to such ranger, constable, or person his license, and the contents of his creel or bag, and the bait used by him for taking, catching, or killing trout.

8. Every trout not exceeding eight inches in length from nose to tip of tail taken or caught by any person shall immediately be returned alive into the water from which the same is taken.

9. No person shall put, throw, or place, or allow to be put, thrown, or placed, in any of the said waters, any sawdust or sawmill refuse, or any thing of any kind or description whatever poisonous, deleterious, or noxious to fish. This shall not apply to *débris* or tailings from mining claims.

10. No person shall take, fish for, catch, or kill in any manner whatever, or have in his possession, any salmon, salmon-parr, or smolts, or the ova, young, or fry of any salmon in any stage whatever; and any of the above-named taken by accident or otherwise shall immediately be returned to the water from whence it was taken.

11. No person other than is provided by the regulations for taking lake trout shall sell, or expose or offer for sale, within the district to which these regulations relate, any of the salmonidæ, trout, perch, or tench, or take, fish for, catch, or kill any of the salmonidæ, trout, perch, or tench in order to make sale of the same.

12. No person shall put, throw, drag, draw, or place, or allow to be put, thrown, dragged, drawn, or placed, for any purpose whatever, any net of any description (except a landing-net) in any of the rivers or streams in the said district, or within half a mile of the mouth or entrance of any such waters, except as may be provided in regulations authorising the taking of lake trout.

13. Licenses to fish for perch and tench only in all the waters of the South Island Eastern Acclimatisation District will be issued by either of

the secretaries of the North Canterbury, Ashburton, South Canterbury, Geraldine, Waitaki - Waimate, Otago, Lakes District, and Southland Acclimatisation Societies, and for any such license a fee of five shillings shall be paid: Provided that it shall not be obligatory upon any of the said Secretaries to issue a license: Provided further that a license to fish for trout shall be deemed to include permission, subject to these regulations, to fish for perch and tench.

14. No person shall take, catch, or kill any perch or tench under six inches in length, nor shall any perch or tench be taken, caught, or killed at all, or had in possession of any person, between the thirtieth day of June in any year and the thirty-first day of January in the following year.

15. The penalty for the breach of any of these regulations shall not be more than fifty pounds.

N.B.—These regulations will possibly be subject to small alterations. Those in force in the North Island are very similar; trout, however, must exceed 9in. to be taken.

INDEX.

In the Index the following abbreviations are used: L., lake; R., river; A., Auckland Acclimatisation District; T., Taranaki or New Plymouth; H., Hawke's Bay; Ta., Tauranga; W., Wellington; N., Nelson; C., Canterbury; Wa., Waitaki and Waimate; O., Otago; S., Southland.

A.

Acclimatisation in New Zealand, 1
Acclimatisation, Success of, in South Island, 31
Acclimatisation, Wellington Province, 4
Acheron R., C., 43
Ahuriri R., Wa., 61
Alpine Char, 8
Alresford, Trout from, 3
Angler, Cautions to, in Fishing Snow-rivers, 16
Angler, Equipment of, Appendix A
Aorere R., N., 38
Aparima R., S., Placing of Salmon in, 8, 12, 81, 83
Artesian Water, 3
Arthur, late Mr., Observations of, 6, 7, 13, 14, 17, 56
Ashburton R., C., 1, 51
Ashley R., C., 44
Auckland Province, 5, 30, 31
Avon R., C., 3, 46

B.

Back Creek., O., 69
Baits, what, allowed, 19
Bait-can, 26
Balclutha, Township of, as Centre, 66
Beaumont Burn, O., 70
Beetle, Use of, as Bait, 21, 32
Black-cleugh Burn, O., 70
Black-fellow, or Creeper, where found, 22
Black-pine, Grub from, 22
Boreal Province, 10
Brogues, 28
Buller R., N., 41
Bullhead, 1, 15
Bullheads, Number of, found in Trout, 15
Bullheads, Use of, as Bait, 25, 35
Butels Creek, O., Yearly Growth of Trout in, 14

C.

Camp-bed, Usefulness of, 30
Camping-out, 29, 45, 50
Camping-out, Outfit for, Appendix
Canterbury, Acclimatisation in, 3
Carpione Trout, 6, 8
Casts, what, required, 20, 22, 46
Catlin's River, S., 73
Centre Hill Creek, S., 77
Char, American Brook, 3, 5
Chatham Islands, Drift of Current towards, 10
Chili, Haplochiton found in, 6
Christchurch, Hatchery at, 3, 8
Cicada, Use of, 21, 35, 76, 78
Clarence R., C., 1, 18, 43
Clinton, Hatchery at, 2, 6, 8
Clinton, Centre for Waiwera, 67
Clutha R., O., 1, 2, 63, 67, 68
Creel, Size of, 28
Creeper, where found and used, 22, 32, 44, 47, 73, 85
Crookston Burn, O., 70
Cust R., C., 46

D.

Dart R., O., 70
Deep Stream, Yearly Growth of Trout in, 13, 73
Devon Minnow, Use of, 2, 15, 23, *et passim*
Disgorger, Use of, 28
Dipton Creek, S., 77
Dog Creek, C., 44

Index.

Dome Creek, O., 74
Dunsdale Creek, S., 81

E.

Earnscleugh Burn, O., 70
Electricity, Effect of, on Trout, 26
Ellesmere L., 47

F.

Falkland Islands, Haplochiton found in, 6
Farr, Mr., Salmon ova procured by
Field Newspaper quoted, 82, 48
Fish, Comparison of, on New Zealand Coasts, 10
Fishing-boxes at Lake Ellesmere, 48
Fishing, Free Nature of, 35, 39, 68, 75, 84, 89
Fishing-tackle, Kinds of, Required, 19-24, 34, 35
Five Rivers, S., 77
Flax, New Zealand, 22
Fly-fishing, Largest Trout taken in, 20
Flies, what, required, 21, 35, 46, 58, 62, 85, 88
Fly-rod, what, required, 20
Food-supply, Abnormal Character of, 15, 65
Fruid Burn, O., 70
Fulton's Creek, Yearly Growth of Trout in, 14

G.

Gaff, Use of, in Spinning, 28
Geen's Spiral Minnow, 23
Geraldine County, Acclimatisation in, 51
Glenomaru R., O., 65
Grayling, Native, 6
Greenstone R., O., 70
Grilse, Capture of, in Aparima, 9
Gows Creek, O., 74
Gum-boots, Use of, 28
Günther, Dr., Opinion of, 82
Gut, Attention necessary to keep, 35

H.

Hae-hae-te-moana R., C., 57
Hakateramea R., Wa., 61
Hall's Creek, C., 48
Hanmer Plains, 43
Happy Valley R., N., 38, 42
Hatching-houses, 3
Hatching out, Different Periods of, 3, 6
Hatchery at Clinton, 2

Hawea L., O., 63, 72
Hawea R., O., 71
Hawera, Central Position of, 33
Hawke's Bay, 30, 32
Hedgehope R., S., 61
Henderson's Stream, A., 31
Heriot Burn, O., 70
History of Acclimatisation, 1,
Horokiwi R., N., 37
Hurunui R., C., 1, 44
Hutt County, Sport in, 34
Hutt R., W., 36

I.

Inaha R.. T., 33
Inanga, 2, *et passim*, 25, 35
Inglewood, T., 33
Irthing R., O., 77
Itchen, Hants, 3

J.

Jacobs or Aparima R., S., 8, 9, 12, 81

K.

Kahawai, Abundance of
Kaihiku R., O., 66
Kakanui R., O., 63
Kaiwera or Otaria R., O., 70
Kaitangata Creek, O., 65
Kaiwhata R., W., 4
Katipo (venomous spider) Rarity of, 7
Kawarau R., O., 63, 70
Koan R., O., 64, 66
Kowai R., C., 44
Kuriwao R., O., 2, 3, 67

L.

Lake Heron, Large Trout in, 14
Lake Heron, Spearing Trout in, 14
Landing-net, 28, 34
Lake Trout, Yearly Growth of, 14
Leader R., C., 44
Lee Stream, Yearly Growth of Trout in, 13, 73
Licenses to fish for Trout, 19
Limestone Caves, 40, 84
Leither R., O., 70
Lindis R., 71
Line, what, necessary, 21, 22
Loch Leven Trout, 3
Lochy R., O., 70
Lora R., S., 79, 81
Lyttelton, Capture of Trout at, 12

M.

Maerewhenua R., Wa., 62
Maitai R., N., 38, 41
Makakahi R., W., 38

Makataura R., T., 34
Makarewa R., S., 77, 80
Manapouri L., 86
Manawatu R., H., 32
Mangawara R., A., 31
Manuherikea R., O., 63
Maoris, Paucity of, in South Island, 31
Mason R., C., 44
Marshall's Creek, 2
Masterton, Hatchery at, 4, 5, 37
Masterton, Fishing at, 22
Mataura R., O., S., 72, 86
Matau R., O., 64
Mavora L., S., 86
Milford Lagoon, 55, 56
Minnow, New Zealand, 25
Minzion Burn, O., 70
Mimihau R., O., 72
Morley R., S., 84
Molyneaux R., O., 63
Morison's Creek, O., 65
Motueka R., N., 38, 40
Motupiko R., N., 38, 41

N.
Native Grayling, 6
New River or Oreti, S., 75
Ngatoros R., T., 33
Ngaruroro R., T., 32
Night-fishing, Method and Character of, 26
North Island, 30

O.
Oamaru Creek, Wa., 62
Ohapi R., C., 55
Ohau L., 2, 61
Okuku R., C., 44
Okoroire R., A., 31
Opihi R., C., 52–54
Opoho, Hatchery at, 6, 8
Opotiki R., C., 54
Opuha R., C., 54
Oraka R., A., 31
Orari R., C., 52
Orawia R., S., 84
Oreti R., S., 75
Orari, Station of, for Winchester, 55
Otago, Acclimatisation in, 2
Otago Harbour, Capture of Trout in, 12
Otaki R., Grayling found in, 6
Otamita R., S., 87
Otapiri R., S., 78
Otautau R., S., 88
Otaria Creek, O., 70
Otemata Creek, 62
Otorohanga R., A., 31

Ova, Transit Experiments with, 5
Ova, Salmon, from England, 8
Owake R., O., 73, 74

P.
Palmerston, 2
Pareora R., C., 61
Pakuratahi R., W., 36
Pegs, Tent, 29
Phantom Minnow, Use of, 2, 15, et passim
Plenty, Bay of, 10, 32
Phormium Tenax (New Zealand Flax), 20
Pilbrow, Mr., Accident to, 16
Piakau R., T., 34
Poles, Tent, 29
Pokaiwhenua R., A., 31
Porirua R. and Inlet, W., 37
Pomahaka R., O., 63, 67
Puerua R., O., 65
Pukaki L., 2, 61
Purakino R., S., 81

R.
Rabbits, Quantity of, in Clarence Valley, 43
Rain-rivers, 2
Rainbow Trout, 5, 31
Rakaia R., C., 1, 2, 24, 48, 51
Rakaia, Accident at Mouth of, 16
Rakaia, Prolific Character of, 49, 55
Rangitata R., C., 1, 51
Rainfall, Small, in Summer, 29
Rankle Burn, O., 69
Rees River, O., 70
Regulations for taking Trout, 19, Appendix B
Requirements for Rain-river, Appendix
Requirements for Snow-river, Appendix
Rhine Salmon, 8
River-beds, Crown Property in, 20
Rivers, Reserve along Banks of, 20
Riwaka R., N., 38, 42
Roding R., N., 39
Rods, Kind of, required, 20
Rotoiti L., 32, 38
Rotorua L., 32
Ruamahanga R., W., 37

S.
Salmon, Californian, 7
Salmon, English, Attempts to Acclimatise, 7, 82
Salmon-fry turned into Aparima, 83
Salmon, Rhine, 8

Scotch Burn Trout, 3
Sea, Division of, into Provinces, 10
Sea-mullet, 2
Selwyn River, Fly-fishing on, 22
Selwyn River, Night-fishing on, 26, 47, 48
Shag, Destructive Nature of, 18
Shag R., O., Yearly Growth of Trout in, 13, 75
Smelts, Number of, 15
Smelts in Trout, 15
Smelts, How to Use, as Bait, 25
Smolt, Difficulties of, on going to Sea, 11
Snakes, Absence of, 29
Snow Rivers, 1, *et passim*
Spin, How to, 24, 25
Spinning, Average Size of Fish taken in, 28
Spinning-baits, Use of, 22, 24, 25, 39
Spinning-baits, Mounting of, 24
Spylaw Burn, O., 70
Staunton R., C., 44
Sutton R., O., 73
Swivels, Use of, 24

T.

Taieri R., O., 72
Takaka R., N., 38
Takapuna L., A., 31
Talla Burn, O., 70
Taranaki, 30, 32
Tararua Mountains, 34, 36
Tarawera L., Ta., 32
Tauherenikau R., W., 37
Taupo L., Ta., 32
Tauranga, Acclimatisation Society of, 82
Tasmania, Trout procured from, 2, 3
Te Anau L., S., 86
Te Apuni, Ta., 33
Tekapo Ferry, C., 61
Temperature of Sea off Otago and Southland, 8, 9, 12
Temuka R., C., 52, 56
Tengawai R., C., 55
Tennyson L., 43
Te Pere-a-tukahio R., Ta., 32
Te Puna R., Ta., 32
Teremakau R., Westland, Grayling found in, 6
Tent, Required Size of, 29, Appendix A
Teviot R., O., 70
Tide, Trout affected by, 17, 50, 68
Traces, Kind of, necessary, 24, 35
Transit of Ova Experiments, 5

Triangles, Mounting of, 24
Trout, Bold Nature of, 18
Trout, Brown, 3, 5
Trout, Carpione, 6
Trout, Capture of at Lyttelton and Other Places in Salt Water, 12
Trout, Capricious Character of, 16, 17
Trout, Cicada-fishing for, 21, 76, 78
Trout, Creeper, Fishing for, 22
Trout, Different Appearance and Size of, 13
Trout, Effect of Electrical Disturbance on, 27
Trout, Food of, 15, 65, 80
Trout, Free Character of Fishing for, 20
Trout, Largest caught with Rod and Line, 14, 18
Trout, Large Takes of, 15, 38, 39, 46, 48, 50, 51, 55, 56, 58, 59, 60, 62, 65, 67, 72, 78, 80
Trout, Licenses to take, 19
Trout, Limit of Size of, allowed to be taken, 19
Trout, Loch Leven, 3
Trout, Position of, in Snow-rivers, 24
Trout, Rainbow, 5
Trout, Reasons for Migratory Habits of, 16
Trout, Regulations for taking, 19
Trout, Result of feeding Pigs with, 15
Trout, Scotch Burn, 3
Trout, Thriving of, in Salt Water, 12, 65, 76
Trout, Sport with, depends on Tide, 12
Trout, Time of Spawning of, 7
Trout, Time of Feeding of, 17, 26
Trout, Yearly Growth of, 13–15
Tuakitoto L., O., 65
Tukituki R., H., 32
Tweed, Salmon-ova from the, 7

U.

Upper Clutha R., O., 71
Upper Taieri R., O., Yearly Growth of Trout in, 13
Upper Thames R., A., 31
Upper Waipa R., A., 31
Upper Waitaki R., C., 62

W.

Waders, Necessity of, 28
Waiau R., C., 1, 44
Waiau R., Ta., 32

Index.

Waiau R., S., 86
Waihi R., C., 51, 58
Waihola L., O., 72
Waihopai R., S., 86
Waikaia R., O., 88
Waikaka R., O., 74
Waikato R., A., 31
Waimakariri R., A., 31
Waimakariri R., C., 1, 2, 45
Waimakariri R., Adventure with Fish in, 28
Waimatuku R., S., 81
Waimea R., S., 88
Waingawa R., W., 87
Waianakarua R., Wa., 63
Wainui R., T., 32
Wainui-o-mata R., W., 36
Wainuioru R., O., 4
Waiohini R., W., 37
Waipa, Upper, R., A., 31
Waingongoro R., T., 33
Waipahi R., O., 69
Waipapa R., T., 32
Waipara R., C., 44
Waipawa R., H., 32
Waipoua R., W., 4, 37
Waimumu R., S., 87

Waiokura R., T., 33
Wairaki R., S., 85
Wairuna R., O., 3, 69
Wairoa R., N., 38
Waitahuna R., O , 66
Waitaki R., C., 1, 2, 24, 61
Waitaki R., Bad Record on, 17
Waitepuka R., T., 34
Wanaka L., O., 63, 72
Wangamoa R., N., 38
Water of Leith R., O., 13, 74
Water, Temperature of, during Hatching-period, 3
Wellington, Acclimatisation in, 4, 5, 34
Whitebait, Quantity of, 15
Whitebait Phantom, Use of, 23, 35, et passim
Wind, Best Kind of, 17, 39, 41
Wind, North-west, Effect of, in Canterbury, 49
Winton Bridge, 79, 80
Winton Creek, 77

Y.

Yearly Growth of Trout, Table of, 13-15

S. PERCY SMITH
Surveyor General

NEW ZEALAND

TROU... 1892

Scale of English Miles

Reference
Chief Towns _____ shown thus DUNEDIN
Minor Towns and Post Offices _____ ANANOA
Roads _____
Railways to 1892 _____

North Island

North Taranaki Bight
NEW PLYMOUTH
Cape Egmont
Mt Egmont

South Taranaki Bight

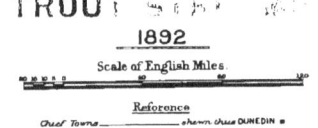

Middle Island

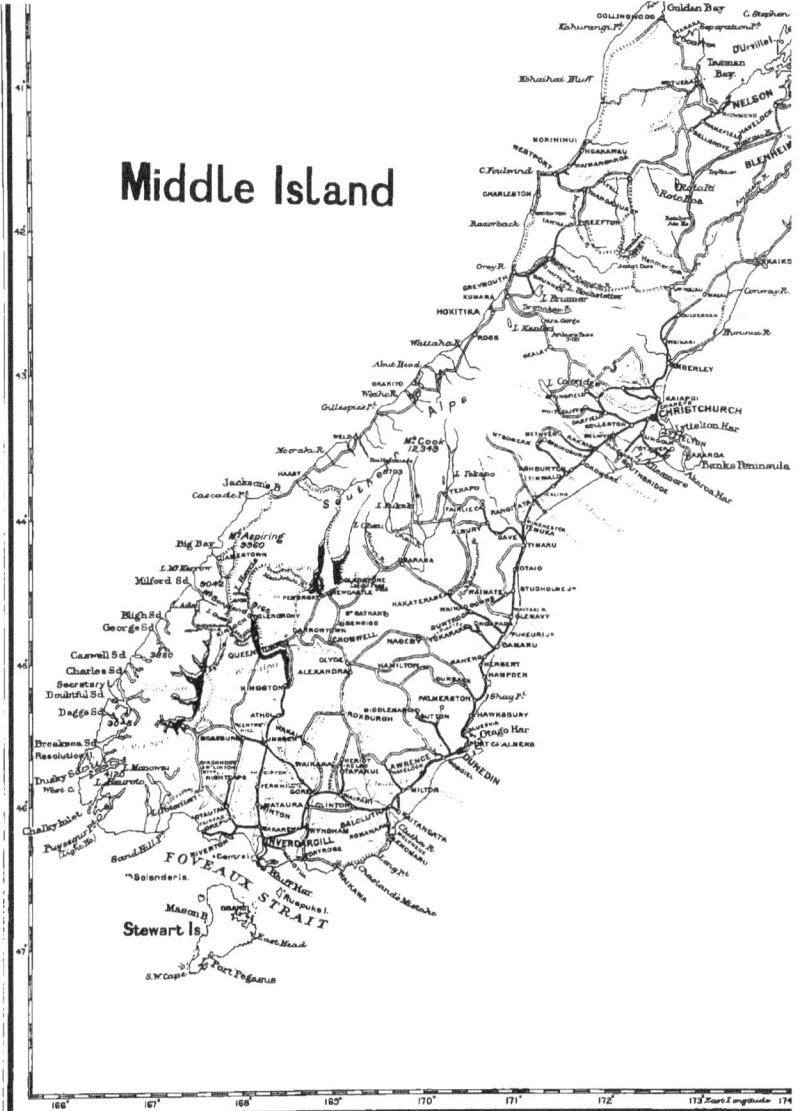

REFERENCE.

NORTH ISLAND.

1 Lake Takapuna
2 Henderson's Stream
3 Oruba Stream
4 Mangawhara Stream
5 Pokaiwhenua Stream
6 Upper Waho River
7 Wainoi Stream
8 Waipapa Stream
9 Te Pona Stream
10 Waiau Stream
11 Eereatuhahia Stream
12 Ngaturo Stream
13 Waijawa River
14 Waingongoro River
15 Inaha Stream
16 Waiokura Stream
17 Manaia Stream
18 Mokotawa Stream
19 Wattupeko Stream
20 Pakau Stream
21 Pakuratahi Stream
22 Waiuniomata River
23 Pouinwa River
24 Bocokiwi River
25 Hururhunga River
26 Waipous River
27 Waingawa River
28 Waiohine River
29 Tauherenikau River
30 Makakahi River

MIDDLE ISLAND.

31 Wangamoa River
32 Wairoa River
33 Takaka River
34 Aorere River
35 Motu. River
36 Riwaka River
37 Ackerson River
38 Rowal River
39 Grey River
40 Okuku River
41 Cust River
42 Avon River
43 Selwyn River
44 Hururata River
45 Irwell Stream
46 Opari River
47 Ohapi River
48 Opihi River
49 Waihi River
50 Pareora River
51 Omaru River
52 Lake Kaitangata
53 Lake Tuakitoto
54 Lovell's Creek
55 Raibiki Creek
56 Waitahuna River
57 Waiwera River
58 Kurtwao Creek
59 Poumahaka River
60 Waipahi River
61 Otaraia River
62 Shotover River
63 Earnscleugh River
64 Beaumont River
65 Fruid Burn
66 Talla Burn
67 Mouton Burn
68 Teriot River
69 Manuherikia River
70 Rees River
71 Lindis Burn
72 Nevas River
73 Gwako River
74 Water of Leith
75 Waitati River
76 Shag River
77 Deep Stream
78 Sutton Stream
79 Lees Stream
80 Mimihau Stream
81 Waikaka River
82 Contre Burn
83 Irthing Stream
84 Dipton Stream
85 Mokareva and Otapiri Rivers
86 Waimatuku River
87 Quaoteu River
88 Orawia and Morley Rivers
89 Mararoa River
90 Waihopai River
91 Owaohia Stream

www.ingramcontent.com/pod-product-compliance
Lightning Source LLC
Chambersburg PA
CBHW031402160426
43196CB00007B/866